ESSENTIALS OF BIOTECHNOLOGY

ESSENTIALS OF BIOTECHNOLOGY

Robert P. Ouellette
VICE PRESIDENT, CORPORATE DEVELOPMENT,
VERSAR, INC., SPRINGFIELD, VA

Paul N. Cheremisinoff
PROFESSOR OF ENVIRONMENTAL ENGINEERING
NEW JERSEY INSTITUTE OF TECHNOLOGY, NEWARK, NJ

TECHNOMIC PUBLISHING CO., INC.
LANCASTER · BASEL

Published in the Western Hemisphere by
Technomic Publishing Company, Inc.
851 New Holland Avenue
Box 3535
Lancaster, Pennsylvania 17604 U.S.A.

Distributed in the Rest of the World by
Technomic Publishing AG

Printed in the United States of America
10 9 8 7 6 5 4 3 2 1

Main entry under title:
Essentials of Biotechnology

A Technomic Publishing Company book
Bibliography: p.
Includes index p. 221

Library of Congress Card No. 85-51484
ISBN No. 87762-437-2

CONTENTS

PREFACE

The outstanding development of biotechnology during the past decade has been a result of science's ability to begin to decipher the codes of life, reprogram the cells and tune molecular switches adopting hosts capable of growing in large quantities. Biotechnology provides the environment on which new bodies of knowledge can expand and act as catalyst for many disciplines.

New technology, equipment and processes will be born in medicine, biochemistry, food, fuels, environmental control as a result of the biological revolution of the last generation. Many of these technologies are complementary and this book views the essentials of biotechnology from a technological viewpoint. Presented is the interaction of several disciplines since all biotechnology is multidisciplinary in nature, challenging the technologist and layman alike.

The reader will find this book useful and informative. Chapters include extensive bibliographies on subject matter presented for further reference and research which will serve as a guide to the more recent literature on biotechnology. Overall this volume should prove useful to both the practitioner as well as the student.

ROBERT P. OUELLETTE
PAUL N. CHEREMISINOFF

1

What is Biotechnology?

INTRODUCTION

Biotechnology is exploding like a Nova, bursting on the everyday scene, challenging the technologist and the layman alike. The field is ancient and new. It is as old as thinking man; and as new as the latest discovery.

It is not possible to review, with scientific rigor, all the findings, advances and discoveries of recent years. We will limit ourselves to a broad sketch on a canvas. We hope that major lines will emerge, guiding our review of what biotechnology has achieved and what it promises to accomplish.

There are no limits to scientific inquiry, still nagging ethical questions persist. We will stay far from such crags and concentrate on the easier topic of science and technology. We display in the following pages the logic of life, the confusing terminology in use today, the prerequisite tools for biotechnological achievements, and some of the emerging concepts.

THE LOGIC OF LIFE

All life is reducible, at the molecular level, to the sequences in DNA (deoxyribonucleic acid). A precise mathematical logic insures that the DNA message is translated into RNA (ribonucleic acid) and that this last code is eventually transcribed into proteins. Proteins serve many functions but none is more important than their role as enzymes catalyzing the entire chemistry of life (Figure 1-1).

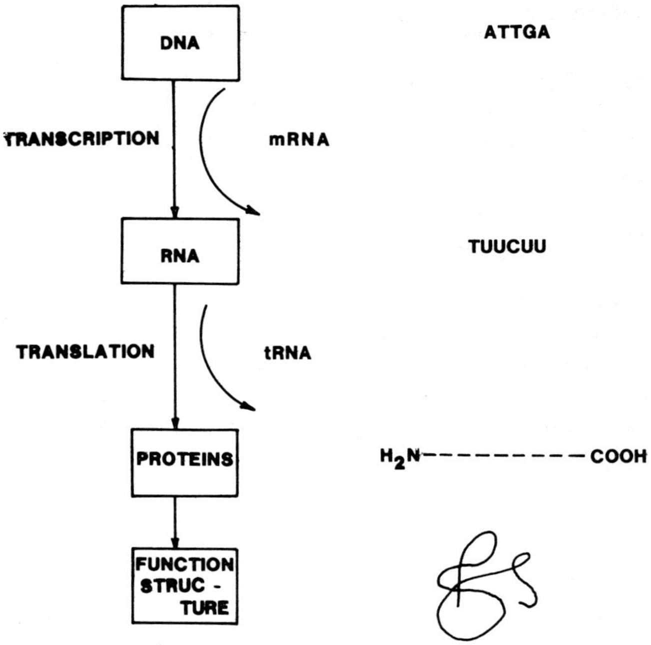

Figure 1-1. The logic of life.

The Central Dogma

A central dogma on the sense and direction of the cellular machinery from DNA to RNA to protein has been proposed. Like all dogma, it sins in its absolute character.

The central dogma of molecular biology states that information transfer never proceeds from protein to nucleic acid. This part of the dogma has never been refuted. The DNA to RNA pathway part of the original formulation had to be revised with the observation that retroviruses transcribe RNA to DNA. The common use of reverse transcriptase in genetic engineering experiments is ample confirmation of the reality of the reverse pathway.

A more modern version of the dogma would be as follows:

Indicating that RNA to DNA information transfers while rare, can occur.

The Building Blocks

Today's engineers might not assemble DNA the way nature did it, but given the basic building blocks, the logic of the assembly is easily recognized. The basic blocks are three: the bases, the sugar moitie, and the phosphate groups.

The Bases

The bases are of five kinds: 2 purines and 3 pyrimidines (Figure 1-2).

The two purines, adenine and guanine, are used both in RNA and DNA; while the pyrimidine thymine is found only in DNA, and uracil appears in RNA. These bases are linked together to form a polynucleotide. The phosphate group of one nucleotide is linked to the deoxyribose group of another, and so forth.

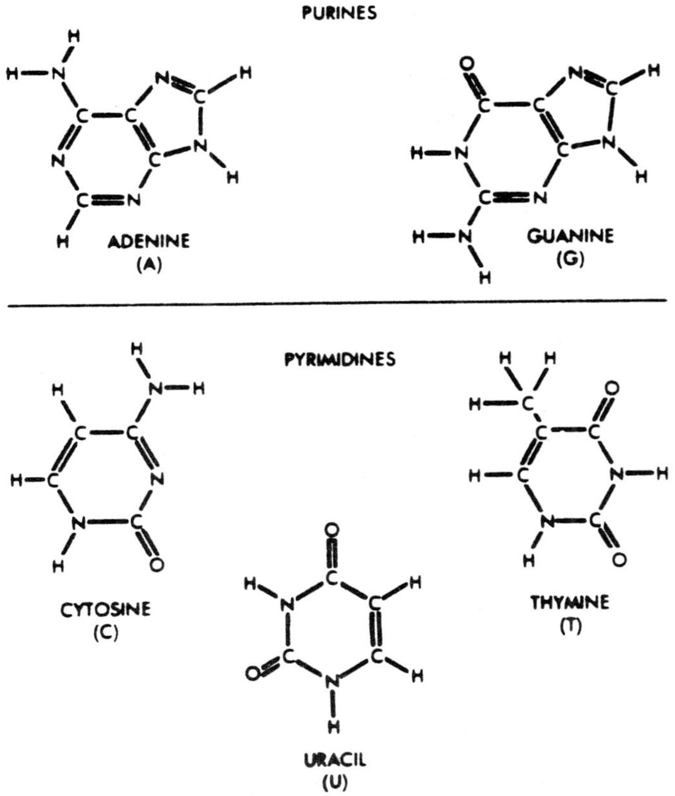

Figure 1-2. The five bases.

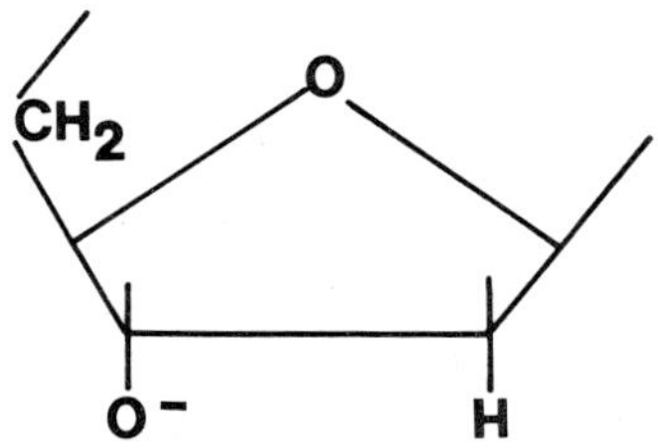

Figure 1-3. The deoxyribose.

The Sugar

The sugar backbone is a deoxyribose (Figure 1-3).

The Phosphodiester Bond

The phosphate group serves as a linkage (Figure 1-4). The linking is from the 5′ carbon of a deoxyribose molecule to the 3′ carbon of a nucleotide (Figure 1-5).

The Double Helix

The double helix of DNA is held together by hydrogen bonds between purines and pyrimidines according to an exact pairing system where A pairs with T and G with C (Figure 1-6).

The double helix structure of DNA was demonstrated in 1953 by Watson and Crick, building on the work of earlier crystallographers [1].

The double helix can now be seen as made up of sugar-phosphate backbones linking the bases. The two strands of the helix are connected by hydrogen bonds (Figure 1-7). The two strands of the DNA molecule have complementary shapes. The strands are twisted in a right-handed double helix. The DNA helix is supercoiled in a compact form. Supercoiling is controlled by enzymes: topoisomerases and gyrases.

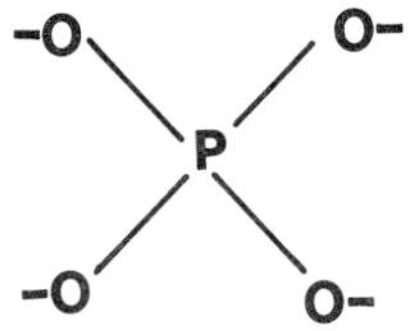

Figure 1-4. The phosphate group.

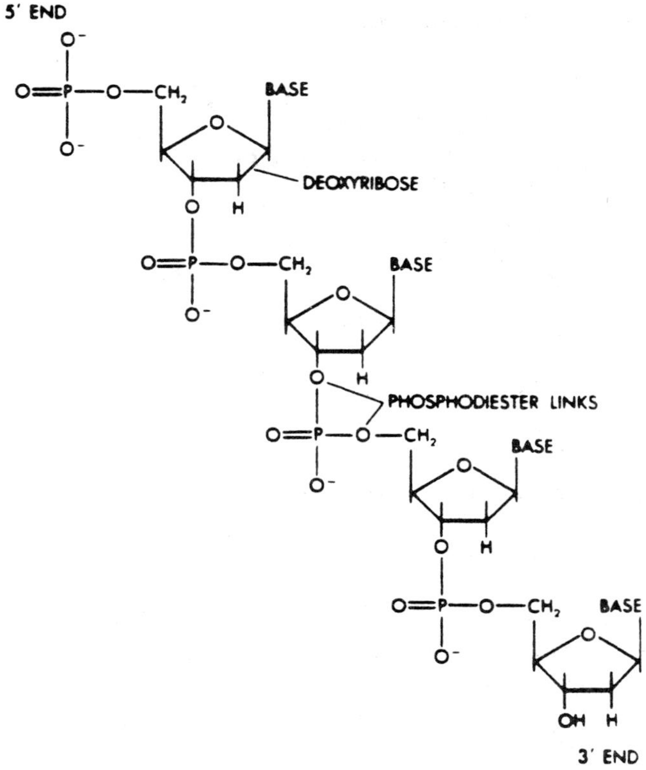

Figure 1-5. Phosphodiester bonds.

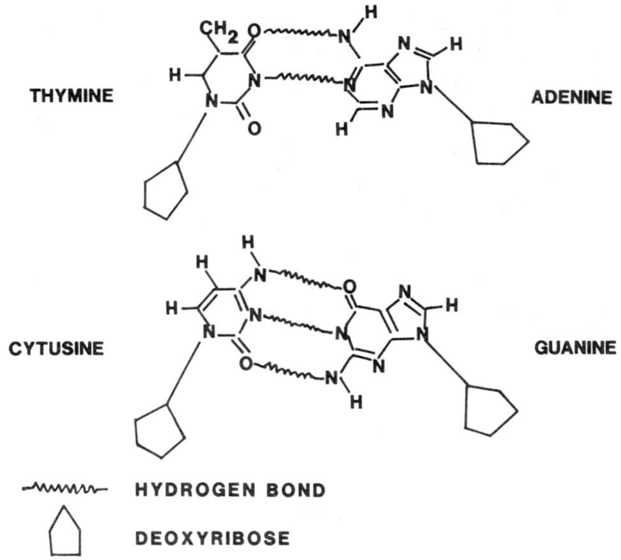

Figure 1-6. Base pairing with hydrogen bonds.

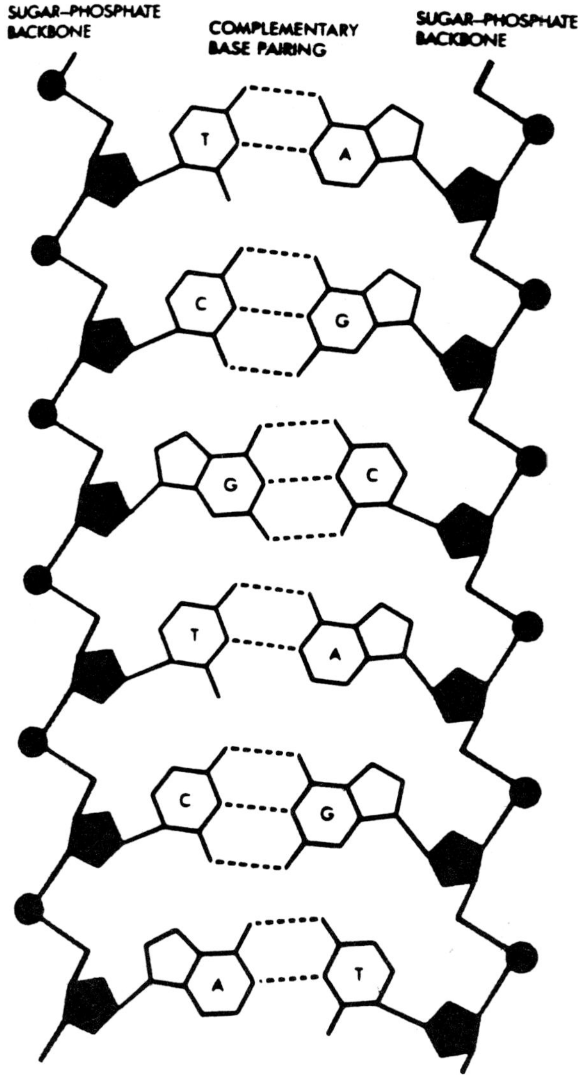

Figure 1-7. The DNA chain.

The DNA Code

We now know how the basic information is stored, transmitted from generation to generation, modified by external agents and noise, and decoded. We know how cells use this information to build structures and condition the functioning of cells, tissues, organs, and organisms.

The basic message is made up of a sequence of four bases: Adenine

(A), Thymine (T) or Uracil (U), Guanine (G), Cytocine (C). The bases are complementary, insuring geometry into the message.

A C
| |
T(U) G

The genetic code is a language based on four nucleic acid bases, through which the 64 triplets specify the 20 amino acids and operators. The DNA code is a coma free, generally non-overlapping degenerate triplet code (Table 1-1).

The Process

The conversion of the DNA code into a protein molecule is a two-step process, involving numerous enzymes.

DNA Duplication

DNA can direct its own replication. The first step in DNA replication is the separation of the DNA double helix. Each strand then serves as a

TABLE 1-1. The DNA Code.

First Position (5′ end)	Second Position				Third Position (3′ end)
	U	C	A	G	
U	PHE	SER	TYR	CYS	U
	PHE	SER	TYR	CYS	C
	LEU	SER	Stop	Stop	A
	LEU	SER	Stop	TRP	G
C	LEU	PRO	HIS	ARG	U
	LEU	PRO	HIS	ARG	C
	LEU	PRO	GLN	ARG	A
	LEU	PRO	GLN	ARG	G
A	ILE	THR	ASN	SER	U
	ILE	THR	ASN	SER	C
	ILE	THR	LYS	ARG	A
	MET	THR	LYS	ARG	G
G	VAL	ALA	ASP	GLY	U
	VAL	ALA	ASP	GLY	C
	VAL	ALA	GLU	GLY	A
	VAL	ALA	GLU	GLY	G

template for the production of two daughters DNA molecules. The separation-replication process goes hand-in-hand.

Transcription

The conversion of DNA to RNA is accomplished by an RNA polymerase. This enzyme binds to the DNA at a correct starting point, then moves rapidly down the DNA chain, adding the correct nucleotides to the RNA chain until the task is completed.

It takes several steps to transform the primary DNA into mature messenger RNA. Certain nucleotides in the chain are modified and others are excerpted by a splicing mechanism, yielding a mRNA chain shorter than the original primary DNA [2].

Translation

The synthesis of protein is accomplished by intracellular particles called ribosomes. The ribosome is a complex apparatus made up of 52 different proteins and three different RNA molecules. It is a quite efficient machine, adding up some 15 amino acids, to the growing chain, per second [3].

The instructions for coding protein are delivered to the ribosome by the messenger RNA (mRNA). Each amino acid is transported to the ribosome synthesis sites by transfer RNA's (tRNA). The tRNA carries an anticodon which matches, in a complementary fashion, the codon of the messenger RNA.

The ribosome has two binding sites, one for positioning the arriving tRNA carrying the selected amino acid, and the other for holding the growing polypeptide chain.

The process we have described above has been compared to a language, to a computer and more generally, to a finite state machine.

The skeleton of a finite state machine can be uncovered by reducing most modern devices for their bare bone essential elements. A mousetrap, a household appliance, a computer, is a finite state machine [4].

At the molecular level, the ribosome and the various transfer RNA's operate as a finite state machine. The input are the four RNA bases (A,U,G,C); the output is a protein made up of infinite rearrangement and ordering of the basic 20 amino acids. The machine must be able to recognize a start signal (AUG) and a stop signal (UAA, UAG, UGA). The machine reads three bases at a time and can perform either of three functions: start the processing, halt, or translate the input into an output (Figure 1-8).

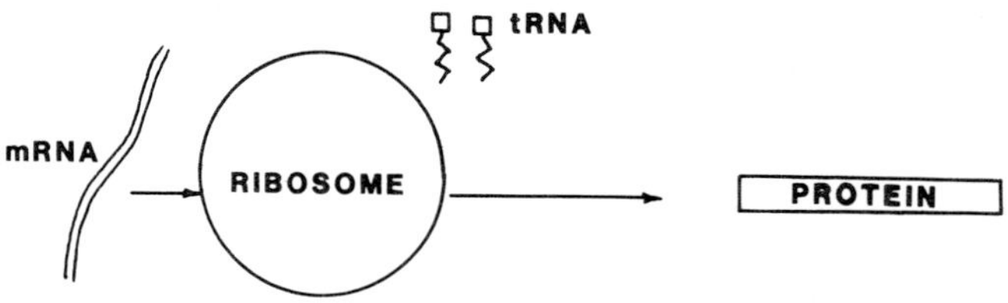

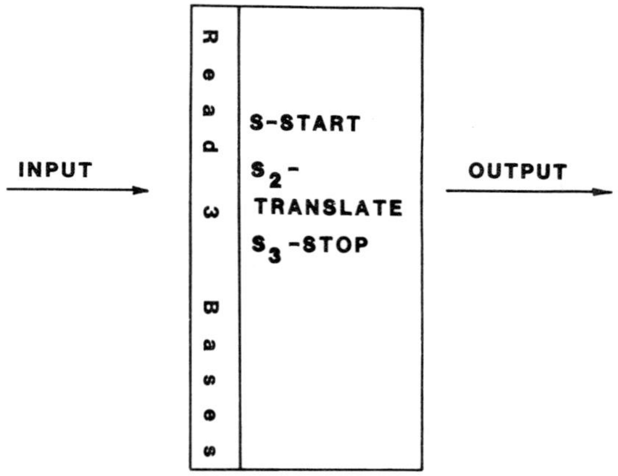

Figure 1-8.

The Protein

Twenty amino acids are commonly occurring in protein.

Glycine	GLY	Lysine	LYS
Alanine	ALA	Arginine	ARG
Valine	VAL	Asparagine	ASN
Isoleucine	ILE	Glutamine	GLN
Leucine	LEU	Cysteine	CYS
Serine	SER	Methionine	MET
Threonine	THR	Tryptophan	TRP
Proline	PRO	Phenylalanine	PHE
Aspartic acid	ASP	Tyrosine	TYR
Glutamic acid	GLU	Histidine	HIS

Proteins are the basis of all living phenomena as structural support, catalyst of equisite selectivity, storage media, transport systems, hormone messengers, etc.

GENETIC ENGINEERING

Terminology

Biotechnology in a certain sense is one of the oldest industries. The production of wine, the manufacturing of cheeses were practiced in the antiquities although many centuries elapsed before a reasonable understanding of these processes surfaced. In this chapter, biotechnology will be used as the broadest term to indicate the use and manipulation of biological materials.

In order to avoid a confusion of language, some terms need to be formally defined.

Biotechnology

This is the broadest term with the most elastic meaning. The term refers to a collection of disparate tools describing the application of biology in technology.

Cloning

A clone is a population of cells or organisms derived from a single cell or an organism by asexual multiplication. Each individual clone has the same genetic makeup.

Genetic Engineering

As the term implies, genetic engineering is the artificial manipulation of genetic material to design a cell or an organism meeting certain preset goals.

Recombinant DNA

The insertion of foreign fragments of DNA in a plant or animal cell is the essential step of all recombinant DNA experiments.

It is often difficult to walk back and forth from the traditional genetic to modern molecular biology. Different levels of resolution, conflicting languages and non-matching units create much confusion.

The Tools

Biotechnology lives by the methods, techniques, processes, equipment and instruments it invents, develops, perfects, uses and utilizes. A variety of tools are required for the conduct of biotechnology experiments:

- Tissue and cell cultures;
- Plasmid vectors
- Host cells;
- Endonucleases.

Two of the most fundamental tools for the practice of biotechnology are plasmids and restriction enzymes.

Plasmids and Cosmids

- Plasmids constitute a menace to modern medicine (by their attack on antibiotics).
- Plasmids are a singular form of environmental protection (by scavenging and destroying foreign molecules).
- Plasmids are the workhorse of biotechnology (by serving as the vehicle to carry gene fragments in recombinant DNA experiments).

Bacterial plasmids are extra chromosomal genetic entity endowed with an autonomous program. They can be isolated as circular supercoiled duplex DNA [5].

In order for plasmid vectors to be successfully used in industrial application, they must be stable. Stability is the ability of transformed cells to maintain plasmids unchanged during their growth. A number of factors affect plasmids stability: growth rate; genetic characteristics of the host cells and the plasmid themselves; environmental stress; etc. [6].

Plasmids are naked fragments of double stranded DNA capable of independent multiplication. They are present in virtually all bacteria. They are generally not essential for the normal metabolism of the host cell. They are only a small fraction (1–3%) of the cell genome. They carry genetic information for conjugation, the efficient use of a variety of nutrients and metabolites and confer resistance to a wide variety of toxic agents. They evolve more or less independently of the host cell. They control their own copy number (from one to more than 100); and are freely exchanged among bacteria by conjugation. They are truly symbionts. They are not basically different than other genes: they code for proteins that are enzymes.

Only a small part of the plasmid is responsible for autonomic behavior: the replication drive unit. This unit is made of the origin (the site for initiation of replication), the initiator substance, the gene controlling the copy number and the genetic function for partition. These units are shown with other genes in the plasmid pI258 [7] (Figure 1-9).

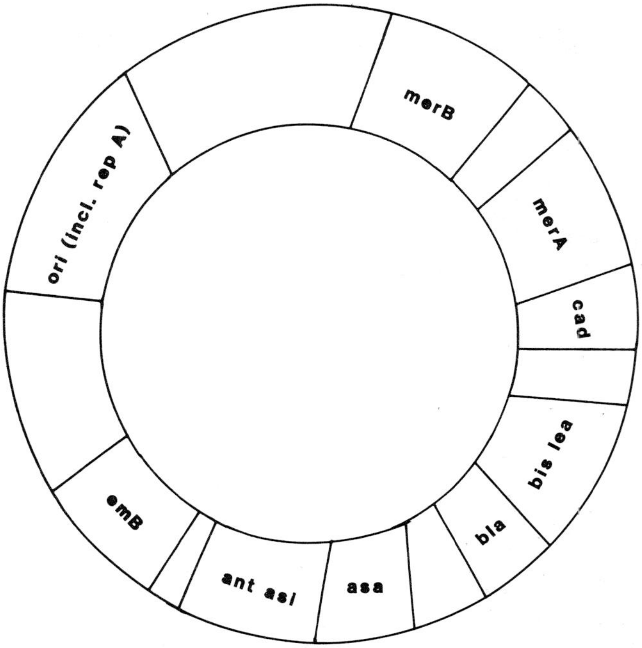

ori = origin
inc = incompatibility
rep A = initiator protein
emB = resistance to erythromycin
ant = resistance to antimony
asi = resistance to arsenite
asa = resistance to arsenate
bla = resistance to penicilin
bis = resistance to bismuth
lea = resistance to lead
cad = resistance to cadmium
merB = resistance to org. mercury
merA = resistance to inorg. mercury

Figure 1-9. General map of pl258

The DNA of plasmids possesses sites recognized by restriction enzymes. It is thus possible to cut open the plasmid ring and to introduce a foreign gene. If this is done with respect to the gene responsible for replication and some resistance genes, the vector can multiply itself and can be controlled by acting on its resistance capabilities. These properties make the plasmid an ideal vector (Figure 1-10).

Cosmids are distinguished from plasmids by the fact that they are aritificially constructed by starting with plasmids and adding a fragment from the chromosome of the bacteriophage lambda called COS (cohesive end site). This site contains the information necessary for the packaging of the phage in the capsid protein. This property allows the cosmid to be included in a virus capsid. The phage recognizes the COS

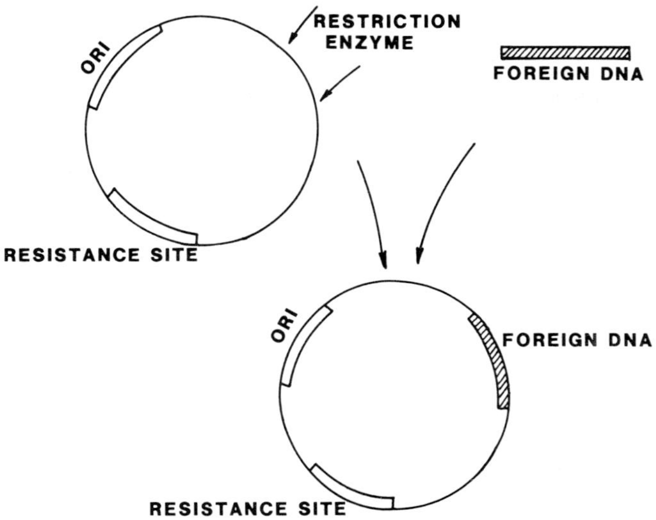

Figure 1-10. Incorporation of foreign DNA in a plasmid.

site and incorporates the DNA into its own genome until it reaches a second COS site. The phage can now infect a bacteria carrying the cosmid into the host bacteria.

The preparation of a cosmid is shown in Figure 1-11 and its incorporation in a host cell is shown in Figure 1-12.

Restriction Enzymes

The purpose to which restriction enzymes are today being widely applied are very different of the role attributed to those enzymes in bacteria. The reason why restriction enzymes evolved in bacteria is not fully understood. It has been proposed that they are defense mechanisms for bacterial cells against unwanted DNA; barriers to breeding between bacteria with some linkage of heterogenous DNA; or as some type of recombination system [8]. Their use as reagents for the analysis of DNA sequences and as tools for in vitro genetic manipulation, is what interests us here.

The type II restriction enzymes recognize target sequences in a double stranded DNA and break the chain within that sequence to give a characteristic set of DNA fragments.

The type II restriction enzymes are especially interesting. These enzymes recognized double stranded DNA sites consisting of 4–6 bases in a twofold rotational symmetry arrangement [9].

Sometimes the restriction enzymes break the phosphodiester bonds

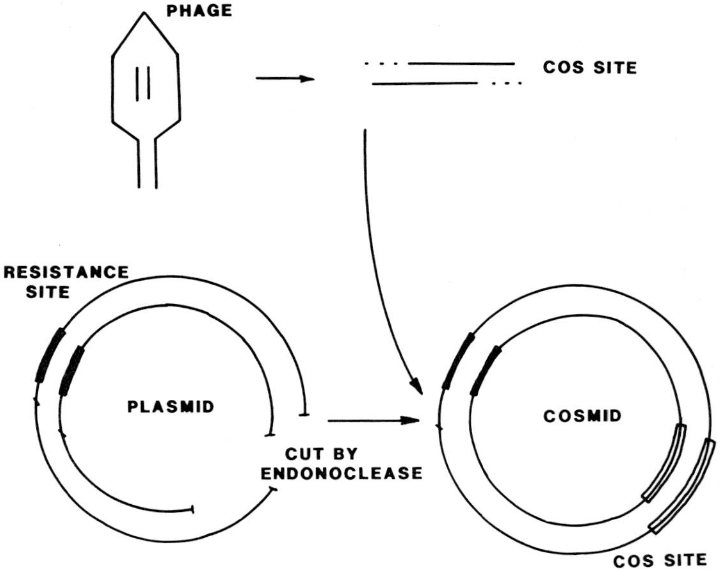

Figure 1-11. Cosmid preparation.

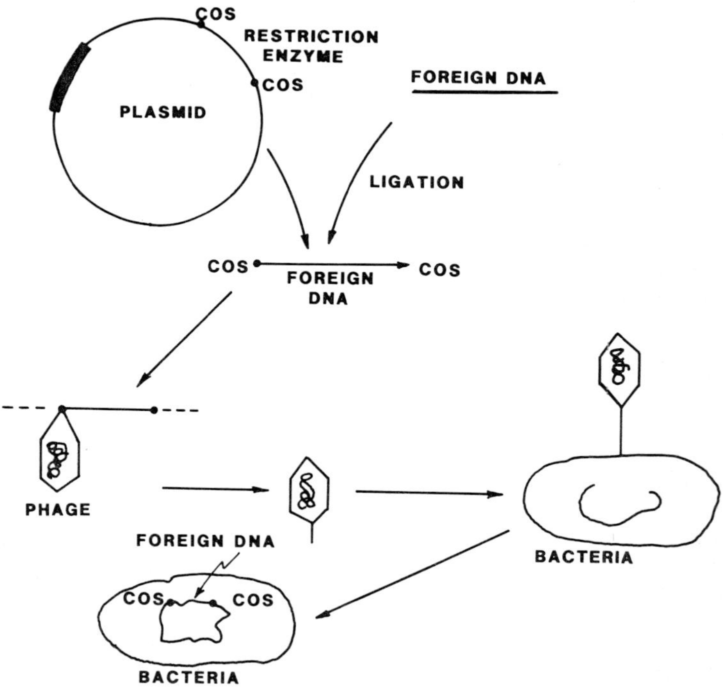

Figure 1-12. Incorporation of cosmid bearing foreign gene into a host bacteria.

between base pairs that are directly opposite giving even breaks or flush-ended breaks. Others break the phosphodiester bonds a few base pairs apart from the target sequence, leaving single stranded projections at either the 3′ or 5′ end. These DNA fragments have the ability to reassociate by normal base pairing between the single-stranded ends of any fragment. Covalent bonds are then easily formed by the action of polynucleotide ligase [8].

The list given in Table 1-2 is a representative set (far from complete) of available restriction enzymes. Table 1-2 identifies the source organism, the target nucleotide sequence for the enzyme, and the position of the phosphodiester bonds broken.

TABLE 1-2. Restriction Endonucleases.

Bacterial Source	Enzyme Name	Target Nucleotide Sequence 5′–3′
Anabaena subcylindrica	AsuI	G↑G-N-C-C
A. variabilis	AvaI	C-Y-C-G-R-G
	AvaII	G↓A-/T-C-C
Arthrobacter luteus	AluI	A-G↓C-T
Bacillus amyoliquefaciens H	BamHI	G↓G-A-T-C-
B. stearothermophilus 1503-4R	BstI	
B. brevis S	BbvS	G-C-(A/T)-G-C
B. globigii	BglII	A↓G-A-T-C-
B. subtilis R	BsuR	G-G↓C-C
Brevibacterium luteum	BluII	
Haemophilus aegyptius	HaeIII	
H. haemoglobinophilus	HhaI	
Providencia alcalifaciens	PalI	
Streptococcus faecalis	SfaI	
Bordetella bronchiseptica	BbrI	A↓A-G-C-T-
Corynebacterium humiferum	ChuI	
H. influenzae,d,b	HindIII,HinbIII	
H. suis	HsuI	
Brevibacterium albidum	BalI	T-G-G-C-C-A
B. umbra	BumI	C-A-G-C-T-G
Corynebacterium humiferum	ChuII	G-T-Y↓R-A-
Haemophilus influenzae d,c	HindII,HincII	
Diplococcus pneumoniae	DpnI,II	↓G-A-T-C
Moraxella bovis	MboI	
M. osleonsis	MosI	
Escherichia coli/RI	EcoRI	G↓A-A-T-T-C
	EcoRI	R-R-A↓T-Y-Y
	EcoRI	↓A-A-T-T
E. coli/R245	EcoRII	↓C-C-(A/T)-G-G

(continued)

TABLE 1-2. (continued).

Bacterial Source	Enzyme Name	Target Nucleotide Sequence 5′-3′
E. coli/PI	EcoPI	A-G-A-T-C-T
H. aegyptius	HaeI	(A/T)-G-G↓C-C-(A/T)
	HaeII	R-G-C-G-C↓Y
H. influenzae H1	Hinh	
Neisseria gonorrhoea	NgoI	
H. aphrophilus	HapII	C↓C-G-G
H. parainfluenzae	HpaII	
Moraxella nonliquefaciens	MnoI	
H. gallinarum	HgaI	G-A-C-G-C 5/10 base pairs
H. haemolyticus	HhaI	G-C-G-C
H. influenzae f	Hinf	G-A-N-T-C
H. parahaemolyticus	HphI	G-C-T-G-A 8/7 base pairs
H. parainfluenzae	HpaI	G-T-T-A-A-C
Moraxella bovis	MboII	G-A-A-G-A- 8/7 base pairs
M. nonliquefaciens	MnlI	C-C-T-C 5/10 base pairs
Providencia stuarti 164	PstI	C-T-G-C-A↓G
Serratia marcescens	SmaI	C-C-C↓G-G-G
Streptomyces albus	SalI	G↓T-C-G-A-C
Thermus aquaticus	TaqI	T↓C-G-A
Xanthomonas badrii	Xba1	T-C-T-A-G-A
X. holcicola	XhoI	C↓T-C-G-A-G
X. papavericola	XpaI	
Xanthomonas badrii	Xba1	T-C-T-A-G-A
X. holcicola	ShoI	C-T-C-G-A-G
X. papavericola	XpaI	
X. malvacearum	XmaI	C-C-C-G-G-G
	XmaII	C-T-G-C-A-G

Source: Ref. 8

Utility Enzymes

Molecular biological experiments cannot be performed without a variety of utility enzymes. Among them the polymerases, ligases and topoisomerases are noteworthy.

Three mammalian DNA polymerases are known, α, β, and γ. Only recently has a clear statement of function emerged [10,11].

- DNA polymerase α is mainly responsible for DNA replication in nuclei.
- DNA polymerase β is involved in nuclear DNA repair.
- DNA polymerase γ is the mitochrondrial replicating enzyme.

The RNA polymerase binds to a promoter site on DNA, unwinds

that part of the double helix and exposes the template bases [10]. The Pribnow region is generally homologous with the sequence TATAATG centered about 10 bases from the initial base of the messenger DNA and is the critical contact region for polymerase-DNA interaction [12]. The termination of transcription is a highly specific process which occurs at a number of well defined terminator sites [13].

A growing family of DNA topoisomerases are gaining better definition. These enzymes convert DNA from one topological form to another (changing the linking number of DNA) and play a profound role in DNA replication, repair, transcription, and recombination [14]. Gyrase is unique in catalyzing the supercoiling of DNA [15,16], giving the DNA duplex its negatively supertwisted form.

Reverse transcriptase allows the synthesis of double stranded cDNA from mRNA. The oligo- dT chain hybridized to the poly-A tail of an mRNA strand serves as a primer for the transcriptase which uses the mRNA as a template for the synthesis of a complementary DNA strand.

DNA ligase is one of many enzymes known to be involved in DNA replication. This enzyme may be used to seal together restriction fragments. Figure 1-13 shows the role of DNA ligase in annealing two DNA fragments with sticky ends.

Therminal transferases are also enzymes of great utility, adding nucleotides to the 3 ′ ends of DNA. This allows creating cohesive ends on blunt-ended DNA fragments.

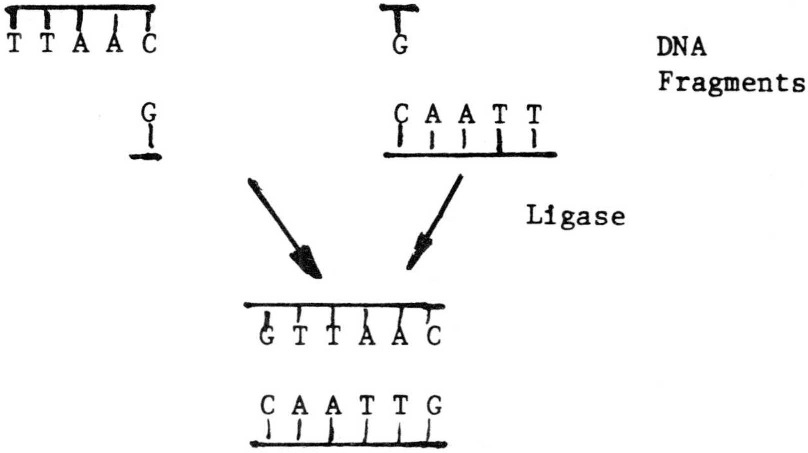

Figure 1-13. Role of ligase.

Recombinant DNA

Recombinant DNA is many things to many people. Rather than describe comprehensively every experiment and its implications, we will give a few practical recipes from which the reader can derive a flavor for the technology or a taste for conducting his own experiment.

Cloning of DNA in a Plasmid

In many cases of genetic engineering, the insertion of an isolated DNA fragment in a plasmid and its subsequent integration in a host cell, is an essential step. This process is displayed in all its apparent simplicity in Figure 1-14.

The Ti Gene Shuttling Machine

The crown gall tumor inducing plasmid (Ti) has proven to be a powerful vector for introducing foreign DNA into plants. The process is shown in a simplified form in Figure 1-15.

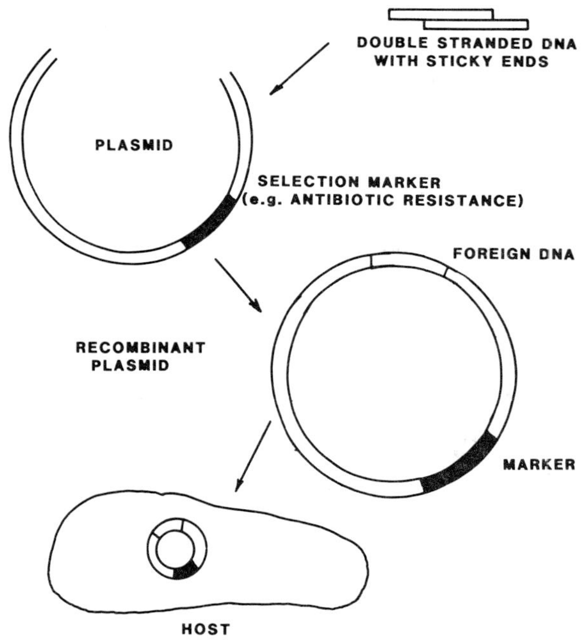

Figure 1-14. DNA cloning in plasmid.

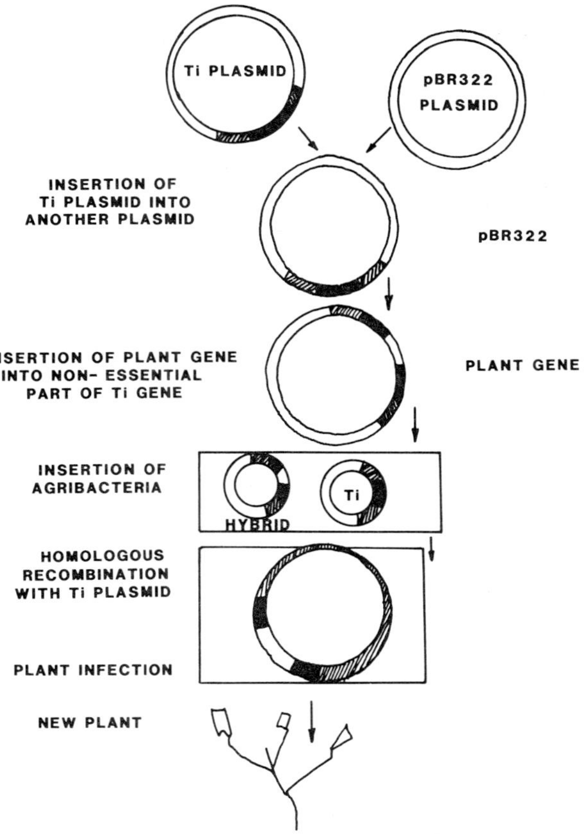

Figure 1-15. Ti gene shuttling machine.

MONUMENTAL AND MOMENTAL ISSUES

Molecular biology is a fertile field, verdant with new developments, feverish with new ideas. In biology, like in many other technical fields, fads and glamourous research areas come and go. The attention and orientation of researchers is sharply focused by social trends. When the interest wanes, we are usually left with a meager crop of new concepts. Some ideas are of the moment only, while others leave a lasting imprint on a technical field or on science as a whole.

Using the well-known crystal ball technique for predicting the future, we review rapidly a few emerging concepts that we believe will last beyond current interest and will dramatically impact on the science and business of biotechnology.

Oncogenes

The word oncogene comes from the Greek, onkos, meaning mass. Oncogenes of animal cancer viruses have cellular counterparts from which they are derived. The cellular gene is not capable of transforming cells into a malignant state, but acquire this ability when they are picked up by viruses.

In the life cycle of viruses, it is not uncommon for them to pick up bits of DNA from the cells they infect. This is the origin of oncogenes, captured by retroviruses from animal host cells.

In a cancerous cell, the viral control element appears to have been inserted by chance into a cellular oncogene. The result is that the oncogene is activated [17], producing vast quantities of protein kinase (probably by point mutation), an enzyme involved in the regulation of cell replication. This enzyme transfers a phosphate group from adenosine triphosphate to amino acids of protein, typically tyrosine. This small change has dramatic effect on cellular biochemistry.

The control of cell growth may originate at the plasma membrane. In cancer cells, adhesion plaques are dismantled, leading to decreased cell adhesion [18,19]. This appears to be the site of action of cellular oncogenes and their enzymes.

Cancer genes would then be interpreted as normal gene of cells driven amuck by changes in their structure.

In summary, three phenomena are associated with the action of oncogenes:

- the production of a protein kinase enzyme
- the phosphorylation of tyrosine residues
- the phosphorylation and inactivation of plasma membranes [20].

Table 1-3 summarizes the extent of our knowledge on oncogenes [21, 22,23]. The table identifies the better known oncogenes, their origin, species of isolation, chromosomal information, and the state of knowledge on protein kinase formation, phosphorylation of tyrosine residues and location on plasma membranes.

Out of these studies of oncogenes, emerges a common pathway to carcinogenesis.

Exon-Intron Organization

The sense of simplicity provided by the crackdown of the genetic code has given rise, a few years ago (in 1979), to a new sense of complexity. The ripple on the calm waters was the discovery that eukaryotic genes are fragmented or split [24] or organized in a mosaic of coding

TABLE 1-3. Oncogenes.

					Oncogene Product	
C-onc	Origin	Species of Isolation	Chromosomal Location	Protein Kinase	Phosphorylates Tyrosine	Located in Plasma Membrane
Src	Rous sarcoma virus	chicken	20	X	X	X
fps-frs	Fujinami sarcoma virus	chicken/cat	15q22-q26	X	X	X
Yes	Y73 sarcoma virus	chicken		X	X	?
ros	UR II avian sarcoma virus	chicken		X	X	Cytoplasmic Membranes
myc	myelocytomatosis virus	chicken	8q24	?	?	?
erb	avian erythroblastosis virus	chicken		?	?	?
myb	avian myeloblastosis virus	chicken	6q15-qter	?	?	?
rel	reticuloendotheliosis virus	turkey		?	?	?
mos	Moloney murine sarcoma virus	mouse	8q22			Cytoplasm
bas	sarcoma	mouse		?	?	?
abl	Abelson murine leukemia virus	mouse	9q34	X	X	X
ras[+]	Harvey sarcoma virus	rat	3	X	?	X
raf	3611 murine sarcoma virus	mouse	3,4	?	?	Cytoplasm
fms	McDonough feline sarcoma virus	cat	Sq34	X	?	Cytoplasm
sis	Simian sarcoma virus	monkey	22q11-qter	?	?	?
fos	FBJ osteosarcoma virus	mouse	2			
ski	avian SKV770 virus	chicken	1			
fgr	Gardner-Rasheed feline sarcoma virus	cat		X	X	?
Ha-ras 1	Harvey murine sarcoma virus	rat	11p15.1-pter	?	?	X
Ki-ras2	Kirsten murine sarcoma virus	rat	12P12-pter	?	?	X
N-ras	neuroblastoma, leukemias, sarcomas	human	1q12-qter	?	?	?
mam	mammary carcinoma	milk, human		?	?	?
neu	neuro-gliobastomas	rat		?	?	?

[+] = multigene family

and non-coding regions [25,26]. The coding regions (Exons) of DNA are interrupted by untranslated segments (Introns). These segments are removed by a splicing system [27].

The Chambon rule holds that bases G-T are the first two nucleotides of any intron, and A-G bases are the last two [24,28].

Some proteins contain many introns. Conalbumin, the egg protein, contains 17 such introns. This is the record so far.

Vertebrates genes have the highest number of introns and are, on the evolutionary time scale, the youngest species [30]. Evidence is accumulating that exons are very old and have been slowly eliminated over the long night of evolution [27,31].

The most likely role from intron is a mechanism for rapid recombination of DNA. The construction of protein from split genes allows for greater flexibility in protein evolution [32,33].

These observations on intron indicate that many new important discoveries remain to be made in molecular biology. While such observations complicate our life, they enrich it as well. Scientists do not know what to make of intron yet. Further research should better explain their role and function now and in the past.

Transposons

Evolution feeds on variety. This variety is brought about by mutations and homologous recombination. Spontaneous mutations are a rare phenomenon and cannot account for the observed exquisite variety. In higher organisms, reshuffling of the genetic material is accomplished by sexual reproduction. In lower organisms, the mechanisms involved include transduction, transformation, and conjugation [34].

To the surprise of biologists, recent observations indicate much fluidity in the genetic material. It appears to be in a constant state of flux [35].

A transposon or transposable element is a nomad bit of DNA capable of moving around in the genome of bacteria or higher organisms. It can insert itself (not at random), and incorporate with itself other DNA fragments, at a large number of sites.

Eukaryotic transposable elements are segments of DNA flanked by sequence repeats, themselves bounded by inverted repeats. This structure is also shared by bacterial transposons with some variations [36]. Some 20 families of mobile dispersed elements are now known [37].

Transposons can join together DNA segments with little homology or ancestral relationships. Indeed, they allow evolutionary small steps, as well as quantum leaps. Their ability to rearrange genetic information and add or delete genetic materials from chromosomes, make them potentially powerful tools in biotechnology.

An interesting question has arisen. Are transposable elements a special form of integrated retroviruses; or are retroviruses a specialized form of transposons? In any event, they share many similarities [36,38].

The interest in transposons goes beyond their contribution to the theory of evolution. They may eventually permit the routine transfer of genes across genetic barriers [39,40,41].

Preproproteins

The process by which proteins pass from one cellular compartment to another and are secreted by cells is not well understood. It has always been assumed that specific membrane receptors recognize proteins to be transported. How does this recognition mechanism work and how do proteins move to the membrane for exportation?

Cytoplasmic proteins are synthesized on free polysomes, while secreted proteins are synthesized on polysomes bound to the endoplasmic reticulum [42].

Many proteins are selected, antibodies, hormones, digestive enzymes, blood clotting factors. These proteins are synthesized in the cytoplasmic compartments. How are proteins due for secretion labelled and recognized and how do specific proteins find their target site? This represents a major logistic protein [43].

David Sabatini and Gunter Blobel proposed an hypothesis to explain the phenomena: the signal peptide hypothesis. This hypothesis has been confirmed in all known cases (with some variant [44]) of both secreted proteins and membrane proteins. Since then much refinement has been added to the hypothesis planting it on firm ground (Figures 1-16 and 1-17).

The common ontogeny of the protein secretion glands speaks for a limited number of common gut proteins as ancestors of the many known secreted proteins [45].

Biological membranes present a diffusional barrier to biological molecules. The transfer of macromolecules, across membranes, from one compartment to another, is an important physiological function. Equally important is the secretion capability of membranes [46].

Mutations in the signal peptide prevent the exportation of protein to the outer membranes [47]. Stringent structural integrity at the cleavage site is required for correct processing [48].

Prokaryotic, as well as eukaryotic signal sequences, share several characteristics.

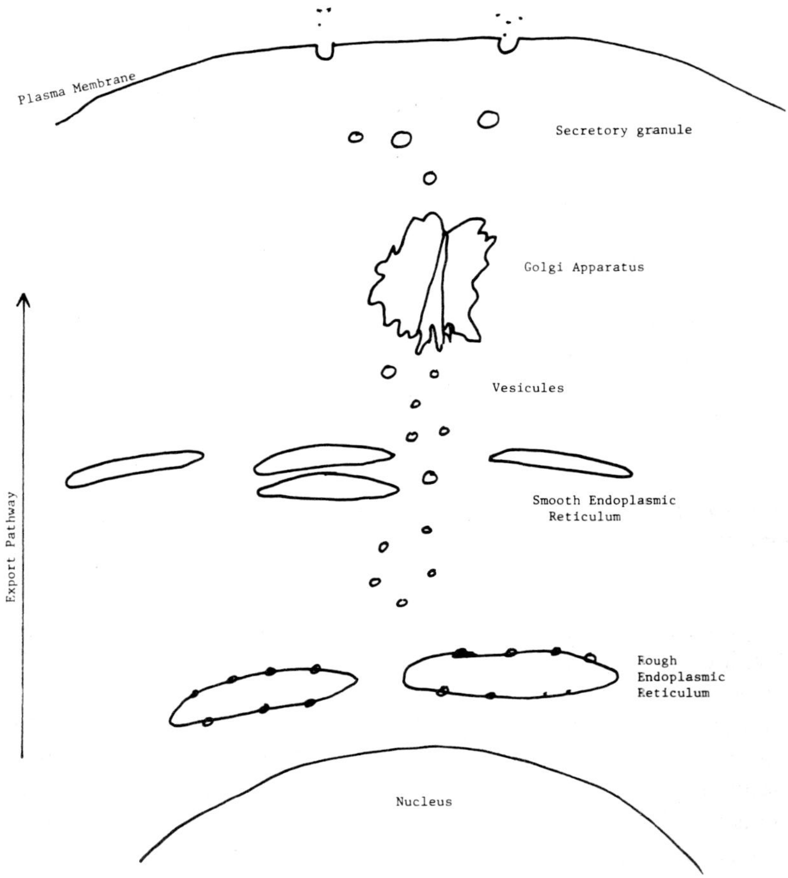

Figure 1-16. Secretory protein pathways.

- short (15–30 amino acids) amino-terminal hydrophobic basic segment followed by a predominantly hydrophobic segment (85–95% of residues) [49]
- one to three basic amino acids (Arg. and/or lys) in the basic portion of the sequence [50]
- the signal segments vary in length and sequences [51]
- the secondary structure of the signal sequence has also been suggested as playing an important role in the initiation of secretion [47,50]
- the signal sequence is cleaved as the protein crosses the membrane [52]

The emergence of a leader peptide from the ribosome is accompanied by a translation block mediated by a cytoplasmic recognition protein

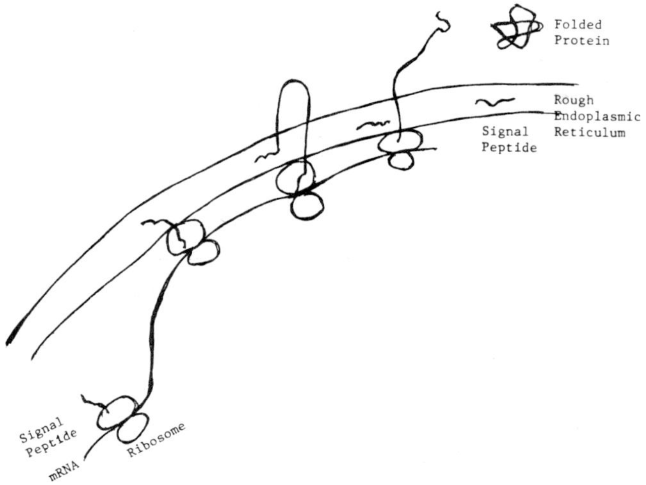

Figure 1-17. Manufacture of secretory proteins by ribosomes attached to the rough endoplasmic reticulum.

complex. This block is released when the complex interacts with a blocking protein, part of the endoplasmic reticulum membrane. The reason for this complex mechanism is to insure a match between the number of nascent protein chains and the availability of pathways for secretion. After release of the block, the synthesis of secretory proteins continues and is coupled to the translocation of the chain across the membrane (Figure 1-18).

Once the processed proteins are transformed into the cysternal side of the reticulum, they moved to the golgi apparatus via small vesicles where they are packaged into secretory capsules.

The signal peptide is removed by signal peptidase, an endoprotease sited on the cisternal side of the membrane [52].

Genetic engineering has not ignored the value of signal peptides as a secretory enhancement mechanism. One of the roles of recombinant DNA is to produce a higher content of specific proteins in bacteria or other hosts. Such proteins are easier to extract if they are secreted from the cell.

When the proinsulin sequence was inserted in the middle of the penicillinase gene (including its 23 amino acid signal sequence), a fused protein was obtained which secretes the penicilinase-insulin hybrid into the periplasmic space [53].

When the NH_2-terminal leader sequence of mouse immunoglobulin light chain precursor was injected into *Xenopus* oocytes, it induced a rapid acceleration in the rate of export of proteins from the oocyte [54].

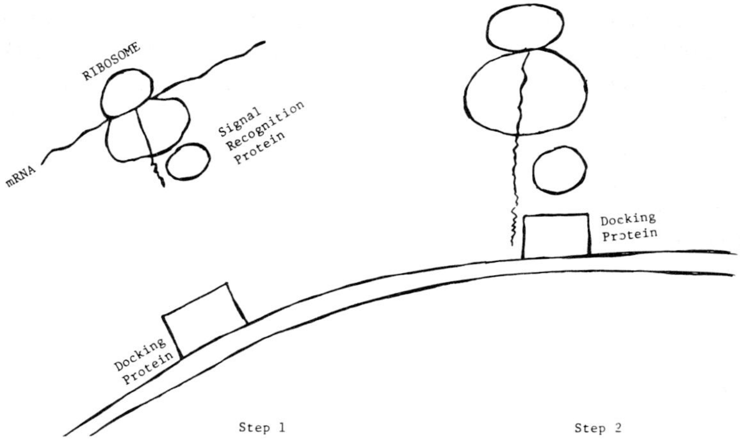

Figure 1-18. Repression and docking.

These examples point out the practical value of using signal peptide to increase the secretion of desired proteins.

Many proteins exist in a proform. That is they must be cleaved proteolytically to release the active hormone. This cleavage is usually accomplished by trypsin-like enzyme. The biologically active domain is released by a tryptic cleavage site (lys-lys, or lys-arg).

Larger precursor inactive forms have been identified for many hormones, proteases, blood coagulation factors and viral proteins. Insulin, lipoprotein, ovalbumin, glucagon, tyrotropin, renin, chorionic somatomammotropin, chymosin, nerve growth factor, albumin, parathyroid hormone, melittin, somatostatin, all exist in a proform prior to activation.

In some cases nature has gone a step further and has packed several active polypeptides in one giant precursor. This is the case for proopiomelanocortin which contains 260 amino acids and is the precursor to adrenocorticotropin (ACTH), α and β melanocyte-stimulating hormones (MSH), β-lipoprotein (β-LPH) and β-endorphin.

The existence of such inactive form of proteins offers interesting possibility for engineering drug delivery systems.

CONCLUSION

Biotechnologies are made possible by the rapid development and utilisation of a variety of techniques, methods, instruments and equipment. High speed centrification, high pressure chromatography, electrophoresis are the tricks of the trade. Monoclonal antibodies, DNA probes, plasmids, vectors are the tools. Protoplast fusion, Ti plasmids, DNA sequencing techniques, gene synthetizing machines, cell culture, fermentation are all part of this arsenal [55].

Biotechnologies will grow in the range of applications promised only if this technical infrastructure continues to grow in parallel, if not ahead of the imagination of scientists.

REFERENCES

1. Watson, J. D. and F. G. C. Crick, "Molecular Structure of Nucleic Acid," *Nature, 171* (4356), 737–738 (1953).
2. Darnell, J. E. J., "The Processing of RNA," *Sci. Am., 249* (4), 90–100 (1983).
3. Nomura, M., "The Control of Ribosome Synthesis," *Sci. Am., 250* (1), 102–114 (1983).
4. Hayes, B., "Computer Recreation," *Sci. Amer., 249,* 19–28 (1983).
5. Hahn, E., "Chemotherapy of Bacterial Plasmids," *Naturwissenschaften, 66,* 555–562 (1977).
6. Imanaka, T. and S. Aiba, "A Perspective on the Application of Genetic Engineering: Stability of Recombinant Plasmids," in *Biochemical Engineering II. A Constratinides,* W. R. Veith and K. Venkatasubramanian, eds., New York Academy of Science (1981).
7. Novick, R. T., "Plasmids," *Scient. Amer., 243,* 103–127 (1980).
8. Murray, K., "Restriction Enzymes and Their Uses in Genetic Engineering," in *Genetic Engineering,* A. M. Chakrabarty, ed., CRC Press, Inc., West Palm Beach, FL (1978).
9. Yuan, R. and D. L. Hamilton, "Restriction and Modification of DNA by a Complex Protein," *Am. Scientist, 70,* 61–69 (1982).
10. Siebenlist, U., "RNA Polymerase Unwinds an 11 Base Pair Segment of a Phage T7 Promoter," *Nature, 179,* 651–652 (1979).
11. Hubscher, U., C. C. Kuenzie, and S. Spadari, "Functional Roles of DNA Polymerases β and γ," *Proc. Natl. Acad. Sci., 36* (5), 2316–2320 (1979).
12. Gilbert, W., "Starting and Stopping Sequences for the RNA Polymerase," in *RNA Polymerase,* R. Lusick and M. Chamberlin, eds., Cold Spring Arbor Lab. (1976).
13. Adhya, S., P. Sarkar, D. Valenzuela, and U. Matra, "Termination of Transcription by *Escherichia coli* RNA Polymerase: Influence of Secondary Structure of RNA Transcripts on ϱ-independent and ϱ-dependent Termination," *Proc. Natl. Acad. Sci., 76* (4), 1613–1617.
14. Wang, J. C., "DNA Topoisomerases," *Sci. Amer., 247,* 94–109 (1982).
15. Fisher, L., "DNA Supercoiling by DNA Gyrase," *Nature, 294,* 607–608 (1981).
16. Cozzarelli, N. R., "DNA Gyrase and the Supercoiling of DNA," *Science, 207,* 953–960 (1981).

17. Jacob, M., "Activation of Some Oncogenes Isolated to Point Mutations," *GEN.*, 34 (March/April 1983).
18. Bishop, J., "Oncogenes," *Sci. Amer.*, 81-92 (March 1982).
19. Abercrombie, M., "Contact Inhibition and Malignancy," *Nature, 284,* 259-262 (1979).
20. Newmark, P., "Tyrosine Phosphorylation and Oncogenes," *Nature, 292,* 15-16 (1981).
21. Land, H., L. Parana, and R. A. Weinberg, "Cellular Oncogenes and Multistep Carcinogenesis," *Science, 222,* 771-778 (1983).
22. Cooper, G. M., "Cellular Transforming Genes," *Science, 218,* 801-806 (1982).
23. Solomon, E. and P. Goodfellow, "Human Gene Mapping Rolls Along," *Nature, 306,* 223-224 (1981).
24. Chambon, P., "Split Genes," *Sci. Amer., 244,* 60-71 (1981).
25. Lewin, R., "The Tree of the Orphanate Genes," *Science, 212,* 530 (1981).
26. Craik, C. S., W. J. Rutter, and R. Fletterick, "Splice Junctions: Associations with Variations in Protein Structure," *Science, 26,* 1125 (1983).
27. Crick, F., "Split Genes and RNA Splicing," *Science, 204,* 264-271 (1979).
28. Mount, S. M., "Sequences That Signal Where to Splice," *Nature, 204,* 309-310 (1983).
29. Fix, J., "Introns are Focus of Gene Research," *C&EN, 23,* 22-23 (July 1979).
30. Heneson, N., "How We Got Our Useless DNA," *New Scientist, 29,* 926 (Sept. 1983).
31. Robertson, M., "Split Genes: The Meaning of Nonsense DNA," *New Scientist, 28,* 262-265 (July 1983).
32. New Scientist, "Human Wandering Genes Can Live on Their Own," *New Scientist, 1,* 18 (April 1983).
33. Blake, C., "Exons—Present from the Beginning," *Nature, 306,* 535-537 (1983).
34. Cohen, S. N. and J. A. Shapiro, "Transposable Genetic Elements," *Sci. Amer., 242,* 40-49 (1980).
35. Lewin, R., "A Broque Turn for Intron Processing," *Science, 218,* 1293-1295 (1981).
36. Flavell, A., "Did Retroviruses Evolve from Transposable Elements?" *Nature, 289,* 10-11 (1981).
37. Mirzabekov, A. D., "Mobile Genetic Elements: New Soviet Studies," *Nature, 291* (1981).
38. Finnegan, D. J., "Retroviruses and Transposable Elements—Which Came First?" *Nature, 302,* 105-106 (1983).
39. Marx, J. L., "A Transposable Element of Maize Emerges," *Science, 219,* 829-830 (1983).
40. Lewin, R., "Can Genes Jump Between Eukaryotic Species?" *Science, 217,* 42-43 (1982).
41. Lewin, R., "Promiscuous DNA Leaps All Barriers," *Science, 219,* 478-479 (1983).
42. Moreno, F., A. V. Fowler, M. Hall, T. J. Silhavy, I. Zabin, and M. E. Schwartz, "A Signal Sequence Is Not Sufficient to Lead β-Galactosidase Out of the Cytoplasm," *Nature, 286,* 356-359 (1980).
43. Freedman, R., "Proteins Have Export Problems Too," *New Scientist, 8,* 376-378 (Feb. 1979).

44. Warren, G., "Stop-Go Proteins," *Nature, 297,* 624–625 (1982).

45. Adelson, J. W., "Enterosecretory Proteins," *Nature, 229,* 321–325 (1971).

46. Blobel, G. and B. Dobberstein, "Transfer of Proteins Across Membranes," *The J. Cell. Biol., 67,* 835–851 (1975).

47. Emr, S. D. and T. J. Silhavy, "Importance of Secondary Structure in the Signal Sequence for Protein Secretion," *Proc. Natl. Acad. Sci., 80,* 4599–4603 (1983).

48. Inouye, S., Chian-Pin Sherman Hsu, and M. Inouye, "Requirements for Signal Peptide Cleavage of *Escherichia coli* Prolipoprotein," *Science, 221,* 59–61 (1983).

49. Palmiter, R. D., J. Gagnon, and K. A. Walsh, "Ovalbumin: A Secreted Protein Without a Transient Hydrophobic Leader Sequence," *Proc. Natl. Acad. Sci., 79* (1), 94–98 (1978).

50. Emr, S. D., J. Hedgpeth, J.-M. Clement, T. J. Silhavy, and M. Hofnung, "Sequence Analysis of Mutations That Prevent Export of λ Receptor, an *Escherichia coli* Outer Membrane Protein," *Nature, 285,* 82–85 (1980).

51. Davis, B. D. and P. C. Tai, "The Mechanism of Protein Secretion Across Membranes," *Nature, 283,* 433–438 (1980).

52. Walter, P., R. C. Jackson, M. M. Marcus, and V. R. Lingappa, "Trypotic Dissection and Reconstitution of Translocation Activity for Nancent Presecretory Protein Across Microsomal Membranes," *Proc. Natl. Acad. Sci., 76* (4), 1795–1799 (1979).

53. Talmadge, K., S. Stahl, and W. Gilbert, "Eukaryotic Signal Sequence Transports Insulin Antigen in *Escherichia coli*," *Proc. Natl. Acad. Sci., 77* (8), 3369–3373 (1980).

54. Koren, R., Y. Burstein, and H. Sureq, "Synthetic Leader Peptide Modulates Secretion of Proteins from Microinserted *Xenopus* Oocytes," *Proc. Natl. Acad. Sci., 80,* 7205–7209 (1983).

55. Cheremisinoff, P. N. and R. P. Ouellette, eds. *Biotechnology, Applications and Research.* Technomic Publishing Co., Inc., Lancaster, PA; Basel, Switz. (1985).

2

Drug Delivery (A Systems Approach)

INTRODUCTION

Drug therapy is the simultaneous solution of an equation involving time, place, and concentration. In other words, we are searching for an optimum delivery rate with minimum side effects. Ideally the drug concentration should reach its effective level immediately, its action should be limited to the desired target sites, and its removal should be total and rapid upon pharmaceutical action. This is an ideal never achieved. For many drugs there is only a fine line between therapeutic action and toxicity level. Not only must we contend with understanding the physical chemistry of the drug, but we must equally master the impact of the formulation on absorption, diffusion, etc.

Drug therapy is entering a new age where it is borrowing much from mathematics, chemistry, physics, engineering and biotechnology. The purpose of this review is to organize, in terms of systems concepts, the state of accumulated knowledge.

THE SYSTEMS

A systems approach to drug delivery allows to dissect the system into its essential components: an active chemical species, a receptor or target, a reservoir or depot, a portal or mode of entry, a force or source of energy, and a rate controlling mechanism. Ultimately a control loop can be added to make the system self sufficient. The elements of the system with their relationships [1–4] are depicted in Figure 2-1. A few

TARGET OR RECEPTOR	PORTAL OR MODE OF ENTRANCE	RESERVOIR	FORCE OR ENERGY SOURCE	RATE CONTROLLING MECHANISM	CHEMICAL SPECIES
Organ	Transdermal	Mechanical	Chemical gradient	Diffusion	Original drug
Tissue	Transmucosal	Liposome	Electrical	Erosion	Procrugs
Antigen/ Antibody	Injection	Mod. cell	Magnetic	Chemical reaction	Modified chemical
	Implantation	Polymer	Osmotic pressure		Immobilization
	Oral	Liquid membrane	Mechanical		Bifunctional antibody
			Thermal		
			Gravity		

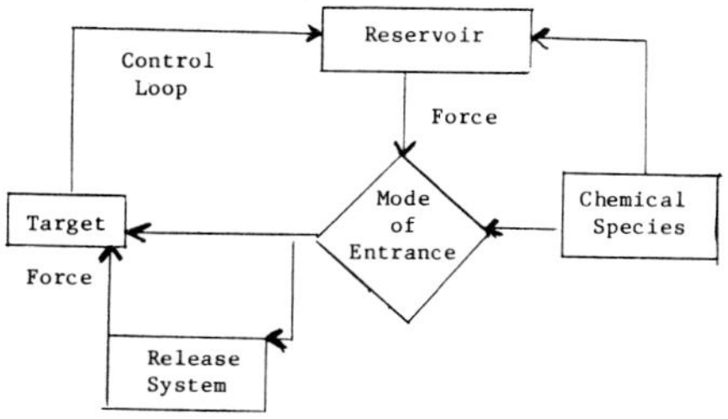

Figure 2-1. A system overview.

examples of drug delivery systems, using the framework defined above, are given in Table 2-1.

THE RECEPTOR

The target for chemical action is usually a tissue, an organ, a group of cells (e.g., cancer), or an invading agent (e.g., virus, bacteria).

The receptor theory of drug action [11,12] has done much to clarify our understanding of drug therapy. Specific binding, molecular shape, thermodynamics, allosteric control sites; all must be taken into account in describing the interaction of drugs with receptors.

While drugs act on living organisms, living organisms also act on the drugs. Drugs in the internal milieu are subject to a variety of attacks intent on degrading them, deactivating them and eliminating them. Many drugs are rapidly filtrated and cleared by the kidney. Others are modified by a number of enzymes or other scavenging molecules. Still others are conjugated with other chemical moities and are rendered inactive. Certain drugs are sequestered in compartments (e.g., liposoluble compounds) where they are no longer available for action on the target.

TABLE 2-1. Examples.

Drug	Reservoir	Portal	Force	Rate Controlling Mechanism	Reference
Many	Bag	Intravenous needle Catheter	Gravity	Drop counter	
Scopolamine	Patch	Transdermal	Concentration gradient	Membrane	5
Insuline	Reservoir	Subcutaneous Peritoneal	Pump	Programmed	6
Pilocarpine	Ocusert	Intraocular	Concentration gradient	Membrane	7,8
Acetazolamide	Tablet	Oral	Osmotic pressure	Orifice	9,10

Still others are rapidly metabolized to inactive products, or products with different actions. It seems surprising that any drug can cause any therapeutic effect.

CHEMICAL SPECIES

The steroid prednisolone can be chemically modified on its 17-ketol side chains. The modified chemical retains its anti-inflammatory activity and its action is local until it enters the systemic circulation where it is rapidly hydrolyzed to inactive steroid acids. The term, antidrug, has been proposed for such therapeutic agents acting locally and without systemic effect.

The modification of original drugs to give them properties such as solubility, or to protect them from inactivation and slow down their release, is not new. Some new modified drugs are right down ingenious. Cefonicid, an antibiotic, contains a sulfonic moities on its imidazole ring. The sulfonic acid binds to serum protein, forming a quasi-reservoir for sustained release into the blood stream.

A prodrug is a drug which is converted to its active form after absorption into the body. The activation process is more often than not enzymatic. The concept recognizes the basic fact that the ideal drug structure for therapeutic action might be far from ideal to insure timely delivery to the target site. The idea is to provide the drug with a transient protective barrier or mask protecting it against inactivation and destructive metabolism, assisting it in crossing biological or biochemical barriers, and minimizing side effects or unwanted effects while the drug is in transit or in general circulation [13].

A few accidentally and intentionally designed prodrugs are well known. Aspirin is the prodrug of salicyclic acid; prontosil is converted in the body to sulphanilamide, levodopa releases dopamine, talampicillin generates ampicillin and the dipivalyl ester of adrenaline secretes the hormone [14].

The brain is well protected from chemical substances. The so-called blood-brain barrier prevents most chemicals from diffusing from the blood vessels to the brain cells. This is a real hindrance to therapy when the drug, like dopamine for the treatment of Parkinsonism, must be delivered directly to the brain.

An ingenious redox delivery system has been designed to work around this problem [15,16]. The drug is combined to a lipid soluble carrier, dihydropyride which helps the drug to cross the physiological barrier. The carrier is then transformed to a charged form which locks the drug carrier within the brain. The drug is then released in the brain, in a sustained manner. Other researchers have taken a different approach, modifying temporarily the barrier to admit molecules to the brain.

PORTAL OR MODE OF ENTRY

The most common mode of entry for drugs is the oral route. In spite of its drawback, it is a quite effective method of administrating drugs. Drugs can be coated to protect the digestive system and to control the release of the active ingredient. Another classic method is the intravenous infusion through a needle or catheter.

A whole family of transmucosal systems, including bronchial, buccal, rectal and sublingual systems, is under active development.

An intrauterine progesterone contraceptive system has shown an efficacy rate of 97%. The polymer membrane reservoir releases the hormone at a constant rate of 65 μg per day for a year [17].

RESERVOIR

The reservoir, depot, or source of the chemical can take a variety of forms: from the traditional bag or bottle for intravenous infusion, to the ubiquitous pills and tablets, to the dreaded syringe, to more advanced patches for transdermal delivery, to specialized devices for intraocular drug delivery. Most interesting are those which have been engineered to deliver their load according to a program or to release their active agent only when meeting the intended receptor.

Pump and reservoir systems embedded under the patient's skin [18] have been used to deliver insulin to people suffering of diabetes [6], and gonadotrophins to infertile women.

Transdermal patches have been used to deliver scopolamine to treat motion sickness and nitroglycerin to treat angina. Many potential applications are being investigated: birth control, treatment of cancer, hypertension, psoriasis, allergies, etc. One of the main advantages of the patch versus oral dosing is that the drugs bypass the liver (where many drugs are destroyed) and enters directly into the bloodstream. There are severe limitations as well, in terms of the quantity of material the patch can hold (10–15 grams/day), and the ability of the drug to be effectively absorbed through a small area of the skin.

FORCE OR ENERGY SOURCE

A source of energy is necessary to insure release of the drug at the target site. This is a pivotal element for constant release or rate control of drug delivery.

Magnetism is one of the forces that can be used to control the rate of drug release. By embedding small metal beads in a polymenic material containing an active drug, a controllable drug delivery system can be turned on and off. Under the influence of an oscillating magnetic field, the metal balls disturb the matrix allowing more drug to be released in the internal milieu. As much as a tenfold increase in drug release was achieved under the influence of weak magnetic fields. The system is being tested for insulin delivery.

A rate controlling device, in the form of a tablet, with a small calibrated orifice, is used for constant rate drug delivery within the gastrointestinal tract. The osmotic pressure rises when water is absorbed through a permeable membrane, and is the force involved in releasing the drug. Acetazolamide has been administered with this device [1].

Electricity or charged ions can be the force in releasing drugs across ion selective membranes or charged membranes.

RATE CONTROLLING MECHANISM

A number of research groups are developing polymer systems for sustained release upon oral administration or subcutaneous implantation [19–21]. For instance, a long acting cough medicine is being tested. Dextrome thirphan, the active ingredient, is trapped inside an ion exchange polymer coated with a semipermeable layer. The pore size of the polymer is the rate controlling mechanism. Water in the body slowly dissolves the drug which slowly must travel down through channels to the surface. Scientists can now design biocompatible, water insoluble, polymer matrix with pores tailored to any desirable size in keeping with the size of the molecule to be trapped and slowly released.

One of the difficulties with such systems is that the release rate is not constant over time. As the drug closer to the surface is released, the medication from deeper in the tablet must travel a longer distance to reach the surface. This problem has been solved in a number of elegant ways based on the ideas of building a pill, layer by layer, with each layer containing a different concentration of the drug [22].

Liquid membranes are rapidly becoming important in medical treatment, in preparing medicinals, in the oxygenation of blood, in the treatment of chronic uremia and in emergency treatment of drug overdose. The membranes can also be used in the reverse mode: a useful drug or enzyme may be encapsulated in the internal phase of the liquid membrane for later slow or "controlled" release of the drug. Successful immobilization of enzymes by liquid for membrane encapsulation has been accomplished with purified phenolase, urease, amylase, lipase, lactase and trypsin. These encapsulated enzymes can be used to facilitate enzyme-induced chemical reactions, in studies of biochemical enzyme mechanisms, in the removal of toxins from the human body, as well as in preparing medicinals.

The medicinals prepared by the use of liquid membranes can potentially be injected and used as traps for toxins present in the gastrointestinal tract, or as slow-release dispensers of drugs or as "reactors." Solid (polymeric) microcapsules are currently used for the above purposes. The problems associated with these microcapsules are that the capsules are prone to swelling followed by rupture, and insoluble microcapsules often lead to fecal compaction. The use of liquid membranes helps alleviate these problems.

Blood oxygenators are devices used to continuously add oxygen to and remove carbon dioxide from blood during cardiovascular surgery, and during the treatment of reversible respiratory insufficiency. These pose a danger in that they eventually lead to damage of blood proteins and red cells. A liquid membrane blood oxygenator has been devised to overcome these problems.

The conventional method of removing the toxins involved in chronic uremia is dialysis. Dialysis treatment is a lengthy, costly procedure and causes a great deal of emotional strain. A liquid membrane system currently being designed for such a purpose can potentially shorten the time required for dialysis by one-third. The underlying mechanism employed in this technique is hydrolysis of urea to carbon dioxide and ammonia by the enzyme urease. The carbon dioxide formed is readily eliminated through the lungs whereas the ammonia is removed by means of a suitably constituted liquid membrane formulation. Such a system requires a dual membrane system—one containing encapsulated urease for controlled release in the gastrointestinal tract, and the second a formulation designed to trap and remove the ammonia formed.

Drug overdose or poisoning treatment methods include peritoneal dialysis and the use of emetics (induced vomiting). An alternate treatment method under development includes the use of liquid membranes. The liquid membranes offer potential advantages such as ease of preparation and administration, good efficacy, minimal patient resistance, and single (one-time) administration. The liquid membranes have been shown to act satisfactorily for phenobarbitol, aspirin and barbituates removal under laboratory conditions. The rate of drug removal is largely dependent on the thickness and viscosity of the membrane as well as the efficiency of the trapping agent.

Intravenously injected liposomes are uptaken mainly in the liver and spleen by macrophages cells of the reticulo-endothelial system. Liposomes, which are nothing more than lipid-like globules filled to the bursting point with chemicals, can be engineered to deliver their therapeutic load in three basic ways.

1. The membrane can leak in the chemicals,
2. The entire liposome can be engulfed by a cell and deliver its dose inside the cell, and
3. The liposome fuses with a cell and adds its drug to the cytoplasm of the cell [23].

The use of liposomes in drug delivery is not without difficulties. The capillaries pores are too small for the liposomes to get through. Colloidal nanoparticles (polymer beads which entrap drugs) suffer from the same problem. Both tend to accumulate in the liver, clogging up the body's disposal system [24].

Liposomes have been relatively effective delivery systems. A new twist has been to target the drug filled liposome by attaching monoclonal antibodies to its outside surface. The system has been tested in treating leukemia. In this way a derivative of methotrexate, highly toxic to cancer and healthy cells as well, was effectively and predominantly targeted to the cancer cells.

Alteration of live blood cells by interchanging some of the natural material for therapeutic agents is another form of drug delivery system [21]. This might be an effective way to bypass the body's natural defense system for up to 20 days (averageable RBC survival time in vivo). One idea being investigated is the use of urease trapped in RBC for the removal of urea in cases of kidney failure.

Neutrophils and some other white blood cells (WBCs) actively sponge up certain drugs. Within the cells, the drugs are hidden and immune to the body's defense system—the antibodies. WBCs actively migrate to sites of infection of areas of damaged cells where they accumulate in large quantities. Scientists are testing the idea of loading WBCs with antibiotics to treat infection or supercharging them with lorcainide to control faulty heart rhythms [25].

Going a step further, a group of scientists has inserted the gene coding for rat preproinsulin in a liposome. Upon insertion, the liposomes are transported to the spleen and liver where they are uptaken by the splenic macrophages and the liver Kupffer cells. The scientists have shown the actual expression of the gene in these organs and the release of the expressed hormone [26].

Along the same line, bifunctional antibodies are being engineered for a variety of targeted drug delivery systems. The drug of interest can be covalently linked to an antibody which will bind specifically to cell surface markers. These bifunctional antibodies are the closest analog to Paul Ehrlich's magic bullet. Another such guided missile is a tandem made of a toxin from diphtheria bacteria and a hormone that bind to specific tumor cells in the pituitary gland [27]. The toxin would be attracted by the tumor cells where the therapeutic hormone would be released.

Still more clever is the idea of immunizing an organism, against certain chemical substances it produces, to achieve modified pharmacological or physiological effects. By combining somatostatin with a human protein and injecting it to the sheep, the animal's immune system was tricked into making antibodies against somatostatin and demobilizing it, giving a chance to natural growth hormone to make its effect felt. The result was substantial body growth (muscle and bone) in these animals. Not only has the immunological feat great industrial importance in the food industry, the technique can be used to control premature puberty in humans. Using the same idea, an analog of luteinizing hormone releasing hormone (LHRH) was used as an alternative to castration. The effect of the vaccine in mice is totally reversible, opening a new path for achieving temporary infertility. Using rats and the same concept, antibodies to glucacon have been produced which might lead to important advances in the treatment of obesity [28].

CONTROL LOOP

With advances in microminitiaturization and microelectronics, it should be possible to add sensors and control logic to a delivery system to close the loop and control the release of drugs based on monitored physiological conditions. Chemical field effect transistors will probably play a key role in this development.

CONCLUSION

Advanced drug delivery systems are not widely used. Their applications have been limited because of technical difficulties limiting their effectiveness and natural resistance to new products.

Their wide distribution and use cannot be avoided. Progresses in membrane technology, minitiaturization, microelectronics, sensors and control systems are all converging on new integrated drug delivery packages. Biotechnology will make a major contribution in monoclonal antibody production, bifunctional antibody design, gene insertion, enzyme therapy, etc. [29].

REFERENCES

1. Goldman, P., "Rate Controlled Drug Delivery," *The New England J. of Medicine, 307* (5), 286–290 (1982).
2. Gregoriadis, G., J. Senior, and A. Truuet, eds. *Targeting of Drugs.* Plenum Publ. Corp. (1982).
3. Zaffaroni, A., "Controlled Therapy is Here," *ChemTech,* 756–761 (Dec. 1976).
4. Chemical Week, "Medication That's Right on Target," *Chemical Week, 11,* 18 (August 1982).
5. Price, N. M., L. G. Schmitt, J. McGuire, J. E. Shaw, and G. Trobough, "Transdermal Scopolamine in the Prevention of Motion Sickness at Sea," *Clin. Pharmacol. Ther., 29,* 414–419 (1981).
6. Albisser, A. M. and W. J. Spencer, "Electronics and The Diabetes," *IEEE Trans. Biom. Engineering, BME-29* (4), 239–247 (1982).
7. Heilmann, K., "Therapeutic Systems, Pattern-Specific Drug Delivery: Concept and Development," *ChemTech,* 1982 (Feb. 1980).
8. Urquhart, J., "Case Study 2: Development of Ocusert® Pilocarpine Ocular Therapeutic System: A Case History in Ophthalmic Drug Delivery Systems," American Pharmaceutical Association, Wash., D.C., pp. 105–108 (1979).
9. Theeuwes, F., "Drug Delivery Systems," *Pharmacol. Ther., 13,* 149–191 (1981).
10. Theeuwes, F., W. Byrne, and V. McGuire, "Gastrointestinal Therapeutic System for Acetazolamide: Efficacy and Side Effects," *Arch. Ophthamol, 96,* 2219–2221 (1978).
11. Sadavongvivad, C., "Receptor Theory of Drug Action," *J. Sci. Soc. Thailand, 3,* 61–72 (1977).
12. Belleau, B., "Receptor Mechanisms and Biochemical Rationales," in *Drug Discovery Advances in Chemistry Series No. 108,* American Chemical Society, Wash., D.C. (1971).
13. McClure, D. A., "Prodrugs as Novel Drug Delivery Systems," *ACS Symposium Series No. 14,* American Chemical Society, Wash., D.C., pp. 224–235 (1975).
14. Barry, B. W., "Drug Delivery Systems," *ChemTech,* 38–44 (January 1983).
15. Bodor, N. and J. W. Simpkins, "Redox Delivery System for Brain Specific, Sustained Release of Dopamine," *Science, 221,* 65–67 (1983).
16. Rawls, R., "New Methods Let Drugs Past Blood-Brain Barrier," *C&EN, 21,* 24–25 (Dec. 1981).
17. Zaffaroni, A., *Proceedings of the 6th Intrnt. Congress of Pharmacology,* Forssan Kirjapaino Oy: Helsinki, pp. 5–53 (1975).
18. C&EN, "Syringe Pump Injects Hormones Periodically," *C&EN, 28,* 40 (Feb. 1983).

19. Rosen, H. B., J. Chang, G. E. Wnek, R. J. Lindhart, and R. Langer, "Bioerodable Polyanhydrides for Controlled Drug Delivery," *MIT Report,* 4-17-83 (1983).
20. Lee, T. K., T. D. Sokoloski, and G. P. Royer, "Serum Albumin Beads: An Injectable, Biodegradable System for the Sustained Release of Drugs," *Science, 213,* 233–235 (1981).
21. C&EN, "Altered Blood Cells: A Way to Give Drugs," *C&EN, 24,* 21 (Feb. 1975).
22. Chemical Week, "Advances in Sustained Release of Medications," *Chemical Week, 11,* 56–58 (May 1983).
23. Science, "Liposomes: Research Applications Grow," *Science, 199,* 1054–1057 (1978).
24. MacKenzie, D. and S. Yanchinski, "New Packages for Old Pills," *New Scientist* (14 July 1983).
25. Science Digest, "Debugging Blood Cells," *Science Digest, 1983,* 82 (Sept. 1983).
26. Nicolau, C., A. LePape, P. Soriano, F. Fargette, and M. F. Juhel, "In vivo Expression of Rat Insulin after Intravenous Administration of the Liposome-Entrapped Gene for Rat Insulin I.," *Proc. Natl. Acad. Sci., 80,* 1008–1072 (1983).
27. Murphy, J. R. and S. Reichlin, "Thyrotrophin-Releasing Hormone-Diphtheria Toxin-Related Polypeptide Congugates," *J. Biol. Chem.,* 10 (Feb. 1983).
28. Dixon, B., "The New Shepherds of St. Kilda," *Science, 83,* 52–56 (Sept. 1983).
29. Cheremisinoff, P. N. and R. P. Ouellette, eds. *Biotechnology, Applications and Research.* Technomic Publishing Co., Inc., Lancaster, PA; Basel, Switz. (1985).

3

Brain Peptides (The Molecules of Hope)

INTRODUCTION

More than 50 small molecules and some larger peptides have been identified as potential neurotransmitters and brain hormones [1–5]. Their smallness make them attractive to the biotechnologist intent on understanding, modifying, and synthesizing them. The possibility to act directly at the sites controlling behavior in health and disease, has not escaped the biotechnologist [6].

An explosion of brain peptides has become an embarrassment to the biochemist [7,8]. Is it possible to distill principles from the accumulated knowledge? Is it too early to try to harness these polypeptides for modifying physiology and behavior in men and animals? A number of entrepreneurs are not waiting for definitive information to proceed along the path.

PEPTIDE HORMONES

The study of hormones is made difficult by the range of action observed at physiological and pharmaceutical (at these concentrations, they are drugs) concentrations. Hormones are involved in complex interrelations with other hormones and body chemicals. When administered to animals, they display a richness of secondary actions beside affecting the target, and display many side effects [9].

Brain peptides implicated as neurotransmitters and hormones display this richness, diversity and potential. Brain peptides have been associated with pain relief, as promoting sleep, as regulating appetite, as stimulating thirst, as aphrodisiac, as fertility drugs, as contraceptive agents and as affecting body temperature [10]. Some of the better known examples include:

- Bombesin is a possible satiety peptide
- The egg-laying hormone (ELH) controls egg laying in snails [11–13]
- β-endorphin is an analgesic
- Vasoactive intestinal peptide (VIP) lowers blood pressure
- Angiotensin elevates blood pressure
- Cholecystokinin causes contraction of the gall bladder releasing bile into the small intestine
- Dynorphin is a potent opioid peptide
- Bradykinin is involved in pain transmission
- LRH has potential aphrodisiac effects
- Factor S may be a sleep promoting peptide
- Vasopressin has strong natriuretic activity.

Table 3-1 is a practical classification of brain peptides [5,9,14]. Table 3-2 is a handy summary and check list of the better known brain peptides [3,5,15].

TABLE 3-1. Classification of Peptides.

Hypothalamic Releasing Hormone
Thyrotropin releasing hormone (TRH)
Luteinizing hormone-releasing hormone (LHRH)
Somatostatin (Growth hormone release-inhibiting factor)
Gonadotropin releasing hormone
Neurohypophyseal Hormones
Vasopressin
Oxytocin
Neurophysine(s)
Pituitary Peptides
Adrenocorticotropin (ACTH)
β-endorphin
α-melanocyte-stimulating hormone (α-MSH)
Other Pituitary Peptides (Not yet conclusive)
Prolactin
Growth hormone
Luteinizing hormone (LH)
Follicle stimulating hormone (FSH)
Thyrotropin
Gut-Brain Peptides
Vasoactive intestinal peptide (VIP)

(continued)

TABLE 3-1. (continued).

Cholecystokinin (CCK)
Substance P
Neurotensin
Met-enkephalin
Leu-enkephalin
Insuline
Glucagon
Gastrin
Bombesin
Secretin
Others
Angiotensin 11
Bradykinin
Carnosine

TABLE 3-2. Important Peptides.

Peptide	Amino Acids	Function
Vasopressin	8	antidiuretic facilitate learning
Oxytocin	8	milk ejection uterine muscle contraction
Tyrotropin releasing hormone		release of TSH body temperature
Somatostatin	14	inhibit release of somatotropin
Luteinizing hormone releasing hormone (LHRH)		sexual behavior release LH, FSH
Angiotensin 11	8	elicit drinking
Gastrin	13,17,34	secretion of HCl
Cholecystokinin (CCK)	33	control of eating
Thyroid stimulating hormone (TSH)		secretion of thyroxine
Vasoactive intestinal peptide (VIP)	28	excitory neurotransmitter
Substance P	11	excitory neurotransmitter
Neurotensin	13	vasodilatation mediate inflammatory response analgesic
Bombesin	14	satiety
Carnosine	2	olfactory signal
endorphin	31	analgesic

(continued)

TABLE 3-2. (continued).

Peptide	Amino Acids	Function
Met-enkephalin	5	
Leu-enkephalin	5	
Dynorphin	17	
Bradykinin	9	hypertension analgesia
Gonadotropin releasing hormone (GnRH)	10	stimulate pituitary to secrete LH, FSH
Growth hormone releasing factor		control growth hormone release
Prolactin	191	milk secretion progesterone secretion
Adrenocorticotropin	39	maintenance of adrenal cortex function
Follicle stimulating hormone		oogenesis, spermatogenesis
Luteneizing hormone	117	androgens secretion ovulation maturation of follicles
Growth hormone	198	body growth

It is apparent that peptides are synthesized as larger molecules precursors (e.g., pro-opiomelanocortin, neurophysin-vasopressin, etc.) that undergo further processing prior to secretion [16–23].

Proteolytic enzymes (proteases) are important in inactivating and terminating the action of peptides [24]. Exopeptidases (aminopeptidases acting at the N-terminus, and carboxypeptidases acting at the C- terminus) are inactivators of peptides. Exopeptidases are involved in the activation of precursor hormones. Commonly the cleavage site is a pair of basic amino acids [7]. Few of these enzymes have been well characterized.

POSSIBLE USES AND MISUSES

It is interesting to review the better known neuropeptides and to speculate as to how industry could use these potent drugs. It is unlikely that any naturally occurring neuropeptides can be used directly as a drug. These brain peptides are rapidly deactivated, cannot penetrate the blood brain barrier, and in many cases are probably addictive. What they can be is starting material for the imaginative pharmacologist.

Somatostatin

This fifteen amino-acid peptide is produced in the hypothalamus. It is also found in other parts of the central nervous system, the pancreas, and the gut. It is an antagonist of growth hormone, and as such suppresses a number of functions, including nutrient absorption by the gastrointestinal track, blood flow in the gut, the secretion of digestive enzymes and the production of stomach acid [25].

It has been proposed to use somatostatin for the treatment of bleeding ulcers. It is being evaluated for the treatment of severe psoriasis and liver damage, and for the prevention of pancreatic irritation following surgery [26].

A phe^4[somatostatin] analogue has twice the activity of somatostatin toward suppressing growth hormone release. There is little doubt that somatostatin and its analogues will find increasing roles in clinical medicine [27].

Growth Factors

Growth hormone releasing factor is being used to treat congenital dwarfism. It might also find a role in wound healing. Transforming growth factor (TGF) also exhibits healing properties [27].

Gonadotropin Releasing Hormone

This releasing factor from the hypothalamus stimulates the pituitary to release luteinizing hormone and follicle stimulating hormone which in turn cause the sex glands to secrete androgens and estrogens. Interesting uses have been proposed for this hormone: the treatment of infertility and as a birth control substance for both men and women. Analogues of the hormone also increase sexual drive [9,27].

Thymosins

More than 30 peptides have been isolated from the thymus gland: including seven thymosins, thymopoietin, thymulin, thymic humoral factor, and thymic factor X. The thymosins act directly or indirectly on the T- and B-lymphocytes, releasing chemicals such as interferon and other peptide hormones. They have been used in the treatment of congenital immune deficiency diseases. They are being tested in the control of rheumatoid arthritis, and have been successfully used with chemical and radiation therapy in treating cancers [27].

Endogeneous Opioids

The number of opioid peptides discovered and analyzed is growing daily [24,28–32]. Their role in health and disease is being unravelled slowly [33,34]. These compounds are all analgesic. The critical structure for opiate peptide activity is Tyr-Gly-Gly-Phe(X), where X is a hydrophobic amino acid such as methionine or leucine [35].

The general concensus is that there are three subtypes of opiate receptors: μ, δ, k. The main ligands for the μ and δ sites are met^5-enkephalin and leu^5-enkephalin and β-endorphin. Recent studies indicate that dynorphin $_{1-13}$ is a good candidate as the endogeneous ligand for k sites [36,37].

Endorphin

β-endorphin, a 31 amino acid peptide is a potent analgesic part of the polypeptide hormone β-lipoprotein produced in the pituitary gland [30]. A variety of analogues have been constructed and tested [40].

Endorphins are released endogeneously by stress, long distance running, acupuncture, sexual activity, and transes [38].

Enkephalins

Methionine(met)- and leucine (leu)-enkephalins are the best known of the group of neurotransmitters or neuromodulators regulating pain transmission by acting as agonists at opiate receptors [41–44]. The other opioid peptides (β-endorphin, dynorphin, and neo-dynorphin) are c-terminal extensions of either leu- or met-enkephalin. Like in the case of many other peptides, a large precursor has been identified. It contains met- and leu-enkephalin in a 7 to 1 ratio [41].

Dynorphin

Dynorphin is a 17 amino acid opioid peptide. It contains the leu-enkephalin sequence at the amino terminus. Dynorphin-(1–13) is a most potent opioid peptide [45–47].

The endogeneous opioids enkephalins and endorphins have little clinical value. They are rapidly deactivated (enkephalins) by enzymes, and the more enzymatically stable endorphin experiences great difficulties in crossing the blood-brain barrier. Two investigatory pathways have been proposed: 1) the development of inhibitors of the enzymes that inactivate enkephalins at their synapses, and 2) the development of enkephalin analogues resistant to enzyme degradation. The second path has produced two compounds being tested in clinical trials.

The concept that neuropeptides might be involved in mental illness is very old. Few significant and consistent results have been reported. β-endorphin and its derivative, [des-Tyr^1]-endorphin appears to have neuroleptic effects [48]. More careful investigations are required before the question can be finally answered.

Bombesin

This tetradecapeptide first isolated from anuran skin reduces appetite in the rat. This peptide is biologically active in the mammalian gastrointestinal track and in the central nervous system [49]. Like other peptides it displays a spectrum of actions upon infusion or injection; induction of gastrin and cholecystokinin release, induction of prolactin and growth hormone secretion, hypothermia (it might be involved in animal hibernation), and hyperglycemia. One biotechnology company is looking into developing bombesin as a satiety compound [50].

Several other peptides have been implicated in the regulation of feeding behavior. CCK, TRH, and insulin would be satiety factors decreasing food intake; β-endorphin would increase food ingestion.

Angiotensin 11

This peptide causes the constriction of blood vessels feeding the skin and kidneys and the dilatation of blood vessels supplying muscles and brain. The hormone causes the adrenal cortex to increase its production of aldosterone which in turn, stimulates the reabsorption of sodium from urine. It also stimulates the secretion of vasopressin with its associated effects. Thus, angiotensin controls directly and indirectly blood volume, pressure, and sodium content. This peptide brings about spontaneous drinking behavior irrespective of the animal's salt or water balance [32].

Vasopressin

Vasopressin and other peptides affect learning and memory [32]. People given long lasting analogues of vasopressin show enhanced attention and higher performance on memory tests. Vasopressin is known to elevate blood pressure briefly and immediately. This might be part of the explanation for enhanced alertness associated with the peptide.

Other hormones which affect memory (ACTH), visual attention (MSH), and reduce anxiety (MSH) have also been described [48,51–53].

Substance P

This peptide discovered some 50 years ago is an 11 amino acid peptide. It appears to be a transmitter of pain signals. It is also a tentative neurotransmitter meeting most criteria for such a denomination [54–56].

Cholecystokinin (CCK)

Cholecystokinin and gastrin share many structural and functional properties. Their identical C-terminal sequence is responsible for their activity. CCK is produced in a variety of lengths (CCK-4, CCK-8, CCK-33) with different biological activities. Physiological activities of CCK include the regulation of satiation, the modulation of catecholaminergic activity, regulation of hypothalamic peptides, and is a potent releaser of insuline (CCK-4) [57–59].

The Pro-opiomelanocortin(POMC) Complex

The complete sequence of pro-opiomelanocortin has been established. Depending on the cleavage pattern, this precursor may serve as the source of different peptides with different biological activity and function [21–23,60].

- β-endorphin is an analgesic and pain killer
- ACTH acts on the adrenal cortex, releasing cholesterol which is converted to corticosterone and cortisol
- α-MSH and β-MSH causes melanin dispersion in the skin of certain lower invertebrates. Their physiological roles in mammals are unknown. Low doses of α-MSH and β-MSH cause natriuresis.

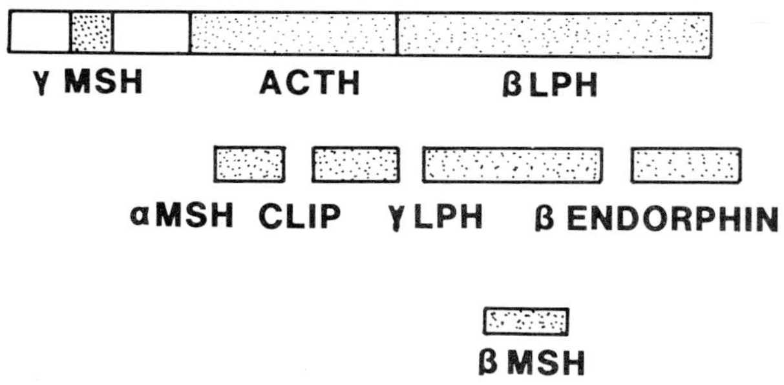

α-MSH and ACTH have been implicated in the alteration of learning behavior
- β-LPH and γ-LPH also possess weak melanin dispersion properties.

The natural concept of linking several proteins into a single vehicle has obvious importance for the biotechnologist. It is an economical method for synthesis, transport, and distribution. The concept is akin to a plumber carrying a tool box from job to job. The presence of several hormones in a large precursor allow to release only the required hormone at the proper site. It also fosters coordinated action among several hormones.

NON-MAMMALIAN PEPTIDES

A large number of interesting peptides have been isolated from a variety of species. Several might be important.

- Urotensin 11 is a somatostatin-like peptide isolated from teleost fishes. It is hypertensive and has smooth muscle and osmoregulatory effects [61].
- Phenypressin[phe^2-arg^8-vasopressin] is found in marsupials [32].
- Physaelemin purified from the skin of the frog, eleidosin purified from the salivary gland of the octopus show amino acid homology with substance P, bombesin and neurotensin [32].
- Reptiles have been shown to possess mesotocin (Ile^8-Oxytocin) and vasotocine (ILe^3-vasopressin).
- In fishes a form of oxytocin, isotocin (Ser^4-Ile^8-oxytocin) has been isolated.

These non-mammalian peptides are of interest. In a certain sense, nature is a scientist and the wild kingdom a vast laboratory. It is difficult to conceive of an experiment not already performed by nature over the years of evolution. Taking advantage of such "experimental" findings would be economical avoiding repeating already performed syntheses and tests.

CONCLUSION

Brain peptides are many and diverse in their function and action sites. This diversity provides both a challenge and an opportunity. The opportunity is to use these fine-tuned chemicals in a variety of medical and industrial applications. The challenge is to unravel the complex interrelations and to develop useful products in the absence of often very basic knowledge.

Fertility drugs, contraceptive agents [62], and analgesics are undergoing clinical trials. More will follow. The more practical and rewarding application might well be on the farm; milk production stimulation and cattle growth enhancement represent large potential financial gains.

The major benefit accrued now is the rapid accumulation of knowledge when both medical scientists and industrialists are simultaneously tackling the problem [63].

REFERENCES

1. Krieger, D. T. and J. B. Martin, "Brain Peptides—Part 1," *New England J. of Medicine, 304* (15), 876–885 (1981).
2. Krieger, D. T. and J. B. Martin, "Brain Peptides—Part 2," *New England J. of Medicine, 304,* (16), 944–951 (1981).
3. Klee, W. A., "Peptides of the Central Nervous System," in *Advances in Protein Chemistry.* Academic Press (1979).
4. Snyder, S. H., "Brain Peptides as Neurotransmitters," *Science, 209,* 976–983 (1980).
5. Schwyzer, R., "Peptides and the New Endocrinology," *Naturwissenschaften, 69,* 15–20 (1982).
6. "The Peptide Link Between Genes and Behavior," *New Scientist, 29,* 303 (July 1982).
7. Newmark, P., "An Embarrassment of Peptides," *Nature, 303,* 655 (1983).
8. Fox, J. L., "Scientists Face Explosion of Brain Components," *C&EN, 19,* 31–33 (Nov. 1979).
9. Parsons, J. A. *Peptide Hormones.* University Park Press, Baltimore (1976).
10. Gurin, J., "Chemical Feelings," *Science, 80,* 28–32 (1980).
11. Heller, E., L. K. Kaczmarek, M. W. Hunkapiller, L. E. Hood, and F. Strumwasser, "Purification and Primary Structure of Two Neuroactive Peptides that Cause Bag Cell Afterdischarge and Egg-Laying in Aplysia," *Proc. Natl. Acad. Sci., 77* (a4), 2328–2332 (1980).
12. "Hormones That Go to the Head," *New Scientist,* p. 455 (19 May, 1983).
13. Lewin, R., "Gene Family Controls a Snail's Egg Laying," *Science, 216,* 720–721 (1982).
14. Hokfelt, T., O. Johansson, A. Ljungdahl, J. M. Lundberg, and M. Schulzberg, "Peptidergic Neurones," *Nature, 284,* 515–521 (1980).
15. Browstein, M. J., J. T. Russell, and H. Gainer, "Synthesis, Transport, and Release of Posterior Pituitary Hormones," *Science, 207,* 373–378.
16. Huang, W.-Y., R. C. C. Chang, A. J. Kastin, D. H. Coy, and A. V. Schally, "Isolation and Structure of Pro-methionine-enkephalin: Potential Enkephalin Precursor from Porcine Hypothalamus," *Proc. Natl. Acad. Sci., 76* (12), 6177–6180 (1979).

17. Brownstein, M. J., "Opioid Peptides: Search for the Precursors," *Nature, 287,* 678–679 (1980).
18. Takagi, H., H. Shiomi, H. Udea, and H. Amano, "A Novel Analgesic Dipeptide from B Ovine Brain is a Possible Met-enkephalin Releaser," *Nature, 282,* 410–412 (1979).
19. Kakidani, H., Y. Furutani, H. Takahashi, M. Noda, Y. Morimoto, T. Hirose, M. Asai, S. Inayama, S. Nakanishi, and S. Numa, "Cloning and Sequence Analysis of cDNA for Procine β-neo-endorphin/dynorphin Precursor," *Nature, 298,* 245–249 (1982).
20. Marx, J. L., "Synthesizing the Opioid Peptides," *Science, 220,* 395–397 (1983).
21. Krieger, D. T. and A. S. Liotta, "Pituitary Hormones in Brain: Where, How, and Why?" *Science, 205,* 366–372 (1979).
22. Liotta, A. S., C. Loudes, J. F. McKelvy, and D. T. Krieger, "Biosynthesis of Precursor Corticotropin/endorphin-, Corticotropin-, α-Melanotropin-, β-Lipotropin-, and β-endorphin-like Material by Cultured Neonatal Rat Hypothalamic Neurons," *Proc. Natl. Acad. Sci., 77* (4), 1880–1884 (1980).
23. Crine, P., F. Gossard, N. G. Seidah, L. Banchette, M. Lis, and M. Chretien, "Concomittant Synthesis of β-endorphin and Melanotropin from Two Forms of Proopiomelanocortin in the Rat Pars Intermedia," *Proc. Natl. Acad. Sci., 76* (10), 5085–5089 (1979).
24. Snyder, S. H., "The Brain's Own Opiates," *C&EN, 28,* 26–35 (Nov. 1977).
25. Bethge, N., F. Diel, and H. Usadel, "Somatostatin—A Regulatory Peptide of Clinical Importance," *J. Clin. Chem. Clin. Biochem., 20,* 603–613 (1982).
26. Alper, J., "Peptides: A Medical Rediscovery," *High Technology,* 60–63 (Sept. 1983).
27. Meyers, C. A., D. H. Coy, W. A. Murphy, T. W. Redding, A. Arimura, and A. V. Schally, "[Phe[4]]Somatostatin: A Potent, Selective Inhibitor of Growth Hormone Release," *Proc. Natl. Acad. Sci., 77* (1), 577–579 (1980).
28. Rossier, J., "Opiod Peptides Have Found Their Roots," *Nature, 298,* 221–222 (1982).
29. Rossier, J. and G. Chapouthier, "Brain Opiates," *Endeavour, 6* (4), 168–176 (1982).
30. Frederickson, R. C. A. and L. E. Geary, "Endogeneous Opiod Peptides: Review of Physiological, Pharmacological and Clinical Aspects," *Progress in Neurobiology, 19,* 19–69 (1982).
31. Henry, J. L., "Circulating Opiods: Possible Physiological Roles in Central Nervous Function," *Neuroscience and Biobehavioral Review, 6,* 229–245 (1982).
32. Bloom, F. E., "Neuropeptides," *Sci. Am., 245,* 148–168 (1981).
33. Marx, J. L., "Brain Opiates in Mental Illness," *Science, 214,* 1013–1015 (1981).
34. Erehart-Treichel, J., "Brain Peptides and Psychopharmacology," *Science News, 110,* 202–203 (1976).
35. Hughes, J., "Opiods Peptides and Their Relatives," *Nature, 278,* 394–395 (1979).
36. Watkins, L. R. and D. J. Mayer, "Organization of Endogeneous Opiate and Nonopiate Pain Control Systems," *Science, 216,* 1185–1192 (1982).
37. Snyder, S., "Opiate Receptors and Internal Opiates," *Sci. Am., 236,* 44–56 (1977).
38. Antelman, S. M. and N. Rowland, "Endogeneous Opiates and Stress-Induced Eating," *Science, 214,* 1149 (1981).
39. Foley, K. M., I. A. Kourides, C. E. Inturrisi, R. F. Kaiko, C. G. Zaroulis, J. B.

Posner, R. W. Houde, and C. H. Li, "β-Endorphin: Analgesic and Hormonal Effects in Humans," *Proc. Natl. Acad. Sci., 76* (10), 5377–5381 (1979).

40. Li, C. H., D. Yamashiro, L.-F. Tseng, W.-C. Chang, and P. Ferrara, "β-Endorphin Ommission Analogs: Dissociation of Immunoreactivity from Other Biological Activities," *Proc. Natl. Acad. Sci., 77* (6), 3211–3214 (1980).
41. Comb, M., P. H. Seeburg, J. Adelman, L. Eiden, and E. Herbert, "Primary Structure of the Hunan Met- and Leu-Enkephalin Precursor and Its mRNA," *Nature, 295,* 663–666 (1982).
42. "Polypeptide from Pituitary Acts as Analgesic," *C&EN,* 26–27 (Nov. 15, 1976).
43. Rossier, J. and G. Chapoutire, "Enkephalines et Endorphines," *La Recherche, 138,* 1296–1305 (1982).
44. Kilpatrick, D. L., A. Wahlstrom, H. W. Lahm, R. Blacher, and S. Udenfriend, "Rimorphin, A Unique, Naturally-Occurring [leu]enkephalin-Containing Peptide Found in Association with Dynorphin and a-neo-endophin," *Proc. Natl. Acad. Sci., 79,* 6480–6483 (1982).
45. Weber, E., C. J. Evans, and J. D. Barchas, "Predominance of the Amino-Terminal Octapeptide Fragment of Dynorphin in Rat Brain Regions," *Nature, 299,* 77–79 (1982).
46. Corbett, A. D., S. J. Paterson, A. T. McKnight, J. Magnan, and H. W. Kosterlitz, "Dynorphin$_{1-8}$ and Dynorphin$_{1-9}$ Are Ligands for the k-Subtype of Opiate Receptor," *Nature, 299,* 79–81 (1982).
47. Goldstein, A., S. Tachibaba, L. I. Lowney, M. Hunkapiller, and L. Hood, "Dynorphin-(1–13), An Extraordinary Potent Opiod Peptide," *Proc. Natl. Acad. Sci., 76* (12), 6666–6670 (1979).
48. Arehart-Treichel, J., "Brain Peptides and Psychopharmacology," *Science News, 10,* 202–203 (1976).
49. Moody, T. W., C. B. Pert, J. Rivier, and M. R. Brown, "Bombesin: Specific Binding to Rat Brain Membranes," *Proc. Natl. Acad. Sci., 75* (11), 5372–5376 (1978).
50. "Bombesin Is Possible Satiety Peptide," *C&EN,* 28 (Nov. 19, 1979).
51. Gitelman, H. J., D. G. Klapper, F. R. Alderman, and W. B. Blythe, "Ala-Gly- and Val-Asp-[Art^8]-Vasopressin: Bovine Storage Forms of Arginine Vasopressin with Natriuretic Activity," *Science, 207,* 893–896 (1980).
52. Land, H., M. Grez, S. Ruppert, H. Schmale, M. Rehbein, D. Richter, and G. Schutz, "Deduced Amino Acid Sequence from the Bovine Oxytocin-Neurophysin 1 Precursor cDNA," *Nature, 302,* 342–344 (1983).
53. Chauvet, M. T., D. Hurpet, J. Chauvet, and R. Acher, "Phenypressin (Phe^2-Arg^8-vasopressin), A New Neurohypophysical Peptide Found in Marsupials," *Nature, 287,* 640–642 (1980).
54. Kessler, J. A., J. E. Adler, and I. B. Black, "Substance P and Somatostatin Regulate Sympathetic Noradrenergic Function," *Science, 221,* 1059–1061 (1983).
55. Marx, J. L., "Brain Peptides: Is Substance P a Transmitter of Pain Signals?" *Science, 205,* 886–889 (1979).
56. "Blowflies Tell the Story of Peptide Evolution," *New Scientist, 15,* 142 (April 1982).
57. Faris, P. L., B. R. Komisaruk, L. R. Watkins, and D. J. Mayer, "Evidence for the Neuropeptide Cholecystokinin as an Antagonist of Opiate Analgesia," *Science, 219,* 310–312 (1983).

58. McCulloch, J., "Peptides and the Microcirculation of Blood Flow in the Brain," *Nature, 304,* 120 (1983).

59. Holst, J. J., S. L. Jensen, and J. S. Morley, "Neural Regulation of Pancreatic Hormone Secretion by the C-terminal Tetrapeptide of CCK," *Nature, 284,* 33–38 (1980).

60. Benjannet, S., N. G. Seidah, R. Routhier, and M. Chretien, "A Novel Human Pituitary Peptide Containing the MSH Sequence," *Nature, 285,* 415–417 (1980).

61. Pearson, D., J. E. Shively, B. R. Clark, I. I. Gerschwind, M. Barkley, R. S. Nishioka, and H. A. Bern, "Urotensin 11: A Somatostatin-like Peptide in the Caudal Neurosecretory System of Fishes," *Proc. Natl. Acad. Sci., 77* (8), 5021–5024 (1980).

62. "Peptide Hormones Tested as Contraceptives," *C&EN,* 18–19 (Feb. 11, 1980).

63. Cheremisinoff, P. N. and R. P. Ouellette, eds. *Biotechnology, Applications and Research.* Technomic Publishing Co., Inc., Lancaster, PA; Basel, Switz. (1985).

4

Chemicals From Plants

INTRODUCTION

A *large number* of drugs and chemical products originate from plants. Some 250,000 species of higher plants are known to man. Less than one percent have been investigated for the production of commercial products.

The purpose of this review is to investigate the state of the art in producing chemicals from plants, especially from plant tissue cultures; and to glance into the near future. The extraction of drugs, chemicals, hydrocarbons, and proteins from plants is reviewed. The role played by plant tissue culture is emphasized.

PLANT EXTRACTS

Plant medicine is today still the norm, rather than the exception, in large parts of the world [1]. In the U.S., one quarter of all prescriptions are for plant-derived compounds. Many chemicals are extracted from plants on a large scale.

Drugs and Chemicals

Many chemicals are extracted from the pulps, juices, stems, roots, and leaves of plants. This is a slow and expensive process. These chemicals are usually complex, difficult to synthesize in the laboratory, and have high value (in excess of $300/lb.). They are mostly drugs (e.g., morphine, codeine), flavorings and fragrances (e.g., spearmint, vanilla, jasmine, chamomile), sweeteners (thaumatin, stevioside), seasonings, specialty enzymes, precursors, and specialty ingredients [2–4]. Tables 4-1 and 4-2 are a compilation of the better known examples of drugs and chemicals commercially extracted from plants [5–7].

TABLE 4-1. Examples of Useful Chemicals Derived from Plants.

Plant	Substances	Uses
gopher plant (*Euphorbia*)	latex sterols	rubber, petroleum substitutes, drugs
jojoba (*Simmondsia*)	long-chain esters	surfactants, emulsifiers, waxes, lubricants, preservatives, cosmetics
buffalo gours (*Cucurbita*)	starch linoleic acid	sweeteners, edible oils
gayule (*Parthenium*)	latex	rubber
scorpion flower (*Phacelia*)	latex chromenes	rubber insecticides
milkweed (*Asclepias*)	latex silk-like fiber	rubber, chemical feedstocks, textiles
juniper (*Pinaceae*)	terpenoids cadinene	antimicrobials insecticides
(*Varthemia candicans*)	sesquiterpene lactones	antimicrobials
(*Jotropha*)	vegetable oils	surfactants
meadowfoam (*Limnanthes*)	fatty acids	surfactants
money plant (*Lunaria*)	fatty acids	surfactants lubricants
bladderpod (*Lesquerella*)	hydroxy fatty acids	lubricants ointments
thisle (*Chamaepeuce*)	hydroxy fatty acids	plastics
kinkaoil ironweed (*Vernonia*)	epoxy fatty acids	plastics coatings
hartleaf christmasbush (*Alchornea*)	epoxy fatty acids	coatings
(*Thaumatococcus*)	thaumatin	sweetener
rubber tree (*Hevea*)	latex	rubber
(*Stevia rebaudiana*)	stevioside	sweetener

Source: modified from Ref. 7.

TABLE 4-2. Examples of Drugs Derived from Plants.

Plant	Substances	Uses
madagascar periwinkle (*Catharanthus roseus*)	leurocristine vincaleukoblastine	leukemia
foxglove	digoxin digitoxin	heart diseases
nightshade (*Atropa belladonna*)	atropine scopolamine	acetylcholine block
Coca (*Erythroxylon coca*)	cocaine	anesthetic
peyote (*Lophopora williamsii*)	mescaline	hallucinogen
ergot	ergotamine	migrane blood vessel constr.
(*Cinchona sp.*)	quinine	malaria
autumn crocus	colchine	gout
(*Strophantus gratus*)	ouabain	heart stimulant
serpent weed (*Rauwolfia serpentina*)	reserpine	hypertension
ipecac (*Uragoga ipecacunha*)	emetine cephaeline	dysentry
(*Chondodentron tomentosum*)	curare	muscle relaxant
mexican yam (*Discorea composita*)	diosgenin	contraceptive
american mandrake (*Podophyllum peltatum*)	podophyllotoxin	tumor
blood root (*Sanguinaria canadensis*)	sanguinarine	cancer
(*Maytenus ovatus*)	maytansine	tumor inhib.
(*Thalictrum dasycarpum*)	thalicarpine	tumor inhib.
(*Euphorbia esula*)	ingenol 3,20-dibenzoate	cancer
(*Coleus*)	forkskolin	glaucoma
caribian sea weed	stypoldione	inhibit cell. division
shikon (*Lithospermum erythrorhizon*)	shikonin	antibacterial anti-inflammatory

Hydrocarbons

Many plants have a high lipid content. The idea of converting algae (e.g., *Nannochloris sp., Chlorella sp., Nonallanthus saliva, Clamydomonas applanate*, etc.) and plants (e.g., *Euphorbia*) into liquid hydrocarbon (18–24 carbons) fuels and chemicals by reaction with hydrogen has been proposed many times [8–11]. Maybe some 1,000 hydrocarbon bearing plants could be used this way. *Euphorbia* appears to be a very good candidate.

Proteins

Proteins can be extracted from seeds, beans, and leafy material.

Leaf protein concentrate (LPC) is a product resulting from the process of green crop fractionation. The first stage involves the maceration of the crop to rupture the cells. This is followed by a pressing process to separate the juice (containing the soluble proteins) from the pressed forage (used for its nutritional value). The proteins in the juice can then be precipitated by acid or heat, and filtered. LPC's show an amino acids profile similar to soybean meal with its deficiency in sulfur containing amino acids. The problem with LPC is its green color, its grassy flavor, and its bitter taste. At present, there is no commercial production of LPC [12].

Full soy flour is produced by cooking and extruding proteins and carbohydrates from the dehulled soybeans. The flour is used in baked products and beverage powders. Soy flour is extracted with hexane solvent. Soy protein isolate is produced from the flour by extracting the proteins.

Much of the developing technology is concentrated in removing objectionable flavors, detoxifying physiologically nocuous components, and providing the necessary texture to the final product.

The market for vegetable protein products is estimated at 2000×10^6 lbs. The primary uses are in animal feed and in baked products [12]. A number of plants currently growing in limited areas could be transplanted to wide areas and become important sources of carbohydrates, proteins, oils, gums, and livestock feed [11].

TISSUE CULTURE

More than 566 culture collections exist today; maintaining bacteria, fungi, algae, protozoa, animal and plant cells lines. These cultures are maintained by subculturing, drying, freeze-drying and cryopreservation [13]. More than 100 laboratories are investigating tissue culture as a method for making useful compounds [14,15].

The Technology

Plant tissue culture is a reasonably well established process. An excised piece of plant tissue is grown in a solid growth medium containing essential sugars, minerals, vitamins, and hormones. The tissue grows into a mass of cell called a callus from which the culture can be maintained indefinitely. If a sample from the callus is placed in a liquid medium under proper agitation, undifferentiated cells separate from the mass, form a suspension, and continue to grow. These cells can often be regenerated into a whole plant by transferring them to the proper medium. Cloned plants from cell culture have been regenerated for tobacco, cabbage, carrot, slash pine, etc.

Plant tissue culture is not as easy as microbial culture:

- the doubling time is very slow, typically 36 to 72 hours
- plant cell culture stop growing as the density increases
- the cells are very large (1000 times larger than microbial cells) and sensitive to stress and strain from stirring and aeration
- plant tissue culture has special light and temperature requirements demanding new types of fermentor
- scale-up has been more difficult, the maximum size achieved is a 20,000 liter fermenter for tobacco cells
- in a few divisions, many plant cells lose their productivity for metabolites of interest
- only batch operations have been successful.

The advantages of plant tissue culture are not negligible:

- it allows to bypass growing whole plants in order to obtain useful products
- it assures a more stable production immune to the vagaries of weather, diseases, and pests
- it insures a uniform product quality
- the supply can be increased to meet demand
- it occupies less space for the same production level.

Secondary metabolites are chemical species produced by plants for which no obvious metabolic role has been identified. Their function appears to be ecological. The search for secondary metabolites is either based on large scale screening of plants (typical is the National Cancer Institute's National Plant Products Program which screened 35,000 plants between 1956 and 1981), the examination of plants based on folklore medicine (periwinkle [*Catharantus*] was analyzed on that basis), and the investigation of plants taxonomically close to plants known to produce pharmaceuticals (this method is currently favored) [15].

Some of the important new directions in tissue culture include:

- immobilization of cells favoring reuse of the cells
- the use of pH change and dimethylsulfoxide to force plants to excrete their metabolites in the medium
- the half-way solution where plant cultures produced precursors are converted to the final product in the laboratory. The 12-β-hydroxylation of digitoxin to digoxin is a good example.

The Market

The timing for commercial production of new chemicals from plants is probably five years. Even there it will be limited to specialized chemicals with value in excess of $300 per pound.

CONCLUSION

Plant cell cultures offer interesting possibilities for the production of specialty chemicals. Few American companies have rushed to meet the challenge. The need for technical breakthroughs in cell culture science and the limited market for most products are not encouraging rapid development.

There is little doubt that plant cultures will replace plant extraction for many drugs, pesticides, fragrances, and seasonings. The future of chemical production by plant tissue culture is more problematic. A few significant technical and commercial successes are needed to give life to this field. A renaissance of interest in natural products is just around the corner. Industry again will take its cue from nature. It is just a question of time [16].

REFERENCES

1. "When the Scientist Meets the Medicine Men," *Nature, 279,* 284–285 (1979).
2. Colwell, R. R., "Biotechnology in the Marine Sciences," *Science, 222,* 222 (1983).

3. Langone, J., "Drugs from the Sea," *Discover,* 60–64 (April 1983).
4. Humphreys, J., "Plants That Bring Health—or Death," *New Scientist, 25,* 513–519 (Feb. 1982).
5. Scott, A., "Biosynthesis of Natural Products," *Science, 184,* 761–764 (1974).
6. "Researchers Get Back to Nature," *Chemical Week,* 28–29 (March 24, 1976).
7. Zaugg, R. H., "Assessment of Future Environmental Trends and Problems: Industrial Use of Applied Genetics and Biotechnology," EPA 600/8-81-020 (1981).
8. "Synthetic Oil: Tapping the Gopherweed," *Chemical Week,* 53–54 (Dec. 12, 1979).
9. "Oil from Algae: Blue Sky or Realistic Goal?" *Chemical Week,* 43–44 (July 18, 1979).
10. "Chemicals Bloom in the Desert," *Chemical Week,* 31–32 (Feb. 7, 1979).
11. Revelle, R., "Flying Beans, Botanical Whales, Jack's Beanstalk, and Other Marvels," *Interciencia, 4* (2), 92–107 (1979).
12. Norton, G. *Plant Proteins.* Butterworths, London (1978).
13. Kirsop, B., "Culture Collections—Their Services to Biotechnology," *Trends in Biotechnology, 1* (1), 4–8 (1983).
14. Graff, G., "Plant Tissue Culture," *High Technology,* 67–74 (Sept./Oct. 1983).
15. Cothif, M. H., "Harvesting Profitable Products from Plant Tissue Culture," *Biotechnology,* 649–657 (Oct. 1983).
16. Cheremisinoff, P. N. and R. P. Ouellette, eds. *Biotechnology, Applications and Research.* Technomic Publishing Co., Inc., Lancaster, PA; Basel, Switz. (1985).

5

Water

INTRODUCTION

Water is such a unique molecule that if it did not exist we would have to invent it. Water is pervasive on earth, essential to life and eccentric in many of its properties [1–3].

Water covers 75% of the surface of the globe. It represents 65 to 70% of the mass of biological material. Life was born in water and few organisms have successfully broken the link to water for significant periods of time. Most biological reactions take place in an aqueous environment. It is not surprising that the major substance a biotechnologist must deal with, is water. Yet we know so little about it.

Some molecules (called amphipathic) have a part soluble in water and another part which is not. Because of the competitive forces involved (attraction-repulsion), these molecules orient themselves with respect to water and form organized structures. Miscelles, and cell membranes are such coherent entities. Proteins are biologically important amphipathic molecules. Of the 20 amino acids, some 25 to 30% are significantly hydrophilic and some 45 to 60% are ionic or contain uncharged hydrophilic side chains [4].

The role of water as a universal solvent is well recognized. The function of water in an oriented form is not as well appreciated and understood.

THE STRUCTURE OF WATER

The structure of water in the form of a cluster, or polymer, or crystal is due to the strength of the hydrogen bonds between water molecules.

Because of these bonds water is highly structured. This structure is disturbed by the addition of any solute. Equilibrium phenomena occurring in aqueous milieu depend on the interaction between water and solute, both in the bulk phase and at interfaces [5].

Water in contact with air tends to minimize its available surface. The ideal shape for energy minimization is a ball or a drop. Water is known to show odd behavior at surfaces, at interfaces or in confined spaces. Liquid-solid and solid-liquid interfaces are very common in biology [6,7].

A number of water anomalies have been reported indicating rapid discontinuity in water properties. These occurrences indicate transitions in the structure of water. Some of the better known anomalies of water include its expansion upon freezing, and the high conductance of ice compared to water [8–11].

A number of models: average uniform models and mixed models have been proposed to explain these anomalies. The uniform model claims that every water molecule behaves like any other water molecule. The mixed models assume at least two species of water [7].

A molecule of water is made up of two atoms of oxygen bound covalently to two atoms of hydrogen. The molecule has a V shape with an angle of 104.5° between the two hydrogen atoms. Each water molecule carries two positive and two negative charges at the corner of a tetrahedron centered on oxygen. When water molecules are grouped together (each water molecule has its four centers associated with four other water molecules through OH bonds) the electrostatic energy is minimized. This arrangement provides large holes between the molecules and weak bonding between hydrogen and oxygen.

The picture of water that emerges is an open structure in constant motion. Large volume and weak cohesion are at the hearth of many peculiar properties of water.

THE PROPERTIES OF WATER

The hydrogen bond is responsible for many of the interesting properties of water. The bent shape of the water molecule can be explained by sp^3 orbitals. Six outer electrons of oxygen and two from hydrogen fill four sp^3 orbitals. Two being bonding orbitals joining oxygen to two hydrogens, and two being non-bonding orbitals containing the lone pairs of the oxygen. The non-bonding orbitals modify somewhat the tetrahedral geometry (leading to a reduced angle between the bonding orbitals (H-O-H angle is 104.5°).

The displacement of the shared bonding electrons toward the more electronegative oxygen and the two pairs of electrons in each of the oxygen's two non-bonding orbitals contribute to pile up negative charges toward the oxygen. The result is that water is the most polar molecule with negative and positive ends. Although a molecule, like water, is neutral, it can be polarized; this means that charges are separated in space or that the centers of gravity of the negative and positive charges do not coincide. The magnitude of the charge separation is the dipole moment (1.85 Debyes for water). The dielectric constant is a measure of the effective polarity of a collection of molecules (78.5 for water). These properties are important since soluble ions interact strongly with dipoles of solvent molecules. Because they are polar, water molecules are attracted to each other and bind through weak (weaker than ionic or covalent bonds) hydrogen bonds.

Water, because of the hydrogen bonds, does not behave like most molecules. According to the rule of the chemical game, water should be a gas at room temperature (what a different world we would live in). Because of its smallness and its important hydrogen bonds, water molecules cling together and form a liquid instead of a gas. Of all the liquids, water has the highest heat of vaporization, because of the large amount of energy needed to rupture the hydrogen bond. Among liquids, water has a low specific heat (17.99 Cal/mol. °), and a high heat of vaporization (10.5 Kcal/mol.).

A surface is a region of high energy. Since at a surface or interface, water molecules have fewer neighbors than in the bulk phase, they have higher energy (the formation of bonds reduces the energy of a system). Chemical systems, like water, tend to minimize energy to insure stability. Hence, they show a strong resistance to increase their surface. This explains why water, here again, has the largest surface tension of all liquids. A high surface tension is the same as a reduced ability to wet a surface. On surfaces, water tends to take the shape of a ball (volume of minimum surface) and does not cover well the surface. Surface active agents (surfactants) weaken the ability of water to minimize its energy and create a weaker, better covered surface.

Because of the structure of water, a number of interesting properties are derived. Water can absorb a great deal of heat without changing its temperature much. Water is a good insulator, but becomes a very good conductor upon addition of salts. Water is a very good solvent especially for ionized species [12]. Water has a very high dielectric constant.

Water can dissociate into its hydrogen (H^+) and hydroxyl (OH^-) ions. Pure water contains very little such ions—only about 10^{-7}. What this means is that it behaves equally well as an acid and as a base.

The interaction energy associated with the H bond of OH groups in water is about 4 Kcal/mole or about ⅔ of the total interaction energy of water.

Water, like other molecules, is in constant motion and is subject to a variety of stress and strain that bend, stretch, and break bonds leading to free OH groups.

Water also displays metastable properties (vitrification, superheating) but these have less interest for the biotechnologist.

WATER AS A SOLVENT

The clustering of water molecules around a solute is called solvation. Water is a very polar solvent and dissolved many chemical species. Water dissolves ionic solids by solvation and polar materials by bonding with the solute (through hydrogen bonds). A molecule without polar group will not be soluble in water.

When two or more liquids or a liquid and a solid are mixed there are often enthalpy and volume changes as a consequence of interaction and reorganization of the species.

Water accepts high quantities of cosolvents only if it forms hydrogen bonds with the new molecules. The same applies to solutes who can either be structure-forming or structure-breaking upon interaction with water.

Water is not a universal solvent but many interesting molecules are soluble, to various degrees, in it. From a biological point of view, water and lipids are the two most important fluids.

WATER IN CHEMICAL REACTIONS

Water is a reaction medium, a transport vehicle, a buffer (for shock, heat), a cooling system, a solvent and a reactant (substrate, product) in chemical reactions. This last role is a most important one. During the formation of a peptide bond (between two amino acids), a molecule of water is eliminated. During photosynthesis (the production of carbohydrates from light, CO_2 and water), water is the source of the oxygen releases during the reaction. Water is a material involved in many biochemical reactions. Chemical reactions involving the formation of water display large enthalpy and free energy changes.

Many biological reactions are controlled by the direct participation of water, its consumption or production, or by catalytic shifts of water attached at or near the site of reactions. For instance, water is often bound to metal ligands of protein, or it is bound to polar groups of large molecules, or acts as a bridge between molecules or fit in clefts, troughs or cavities created by molecular arrangements.

The presence of water can potentially have many dramatic effects on chemical reactivity. It can change conductivity and activation energy, for instance.

INTERACTION OF PROTEINS WITH WATER

Protein crystals are highly hydrated macromolecules. The current thinking is that the protein-water organization includes three kinds of water [13].

a) There are a few molecules of water tightly bound through specific interaction at locations such as active sites, bound to metal ions or in the interior of folded globular proteins. Such water molecules may have become entrapped during folding or are essential to catalytic action. Sixteen molecules are found in the interior of chymotrypsin; nine are associated with the active site of carbonic anhydrase and eight are found within the subtilisin molecule.

b) A larger number of molecules (from 100s to 1000s) are not so tightly bound and make up the "unfreezable water" part of the complex. Some 200 molecules are associated with subtilisin and more than 1000 with myoglobin.

c) A third type of water forms an hydration shell that overlays bound water and acts as a transition with the bulk water of the solution. It is sometimes called transition water.

For many years dissident scientists have maintained that intracellular water is a different structure than liquid water or ice. The details of such structure and its role remain obscure.

The amount of water uptaken by proteins in a 90% relative humidity stabilizes around between 0.2 and 0.6 gram of water per gram of protein. Protein crystals average 43% of solvent by volume. Indeed proteins are well hydrated and this high aqueous solvent concentration has been correlated with their stability.

The number of water molecules needed to form a monomolecular layer of water on a sufficiently large protein (20,000 to 30,000 molecular weight) correspond to about 0.6 g H_2O/g. of protein, or twice the estimated hydration level of most proteins.

Water has often been given a stabilization/destabilization role in DNA and proteins [14]. It is known that water has the ability to inactivate irreversibly DNA and other supramolecular complexes. Water in certain concentrations (greater than 30%) stabilizes the double helix. Polymers in solution go from a water-avid swollen state to a compact squeezed out structure during the denaturation to native form cycle [15].

During protein folding, dramatic changes take place. Even for the smallest polypeptide, a reduction in area of two-thirds or more is achieved between the extended chain and the native folded structure. During the folding process, much water is squeezed out such that the interior of a protein is more like a solid than a liquid [16–18].

Water also competes with peptide groups for hydrogen bonds. In that respect, the role of water can more truly be called regulatory than stabilizing.

No doubt water molecules act as dipoles and interact with macromolecules around them and orient themselves around their charged surface. These water molecules have reduced freedom. NMR analyses have shown that the relaxation time of intracellular water is different than that of free water. The relaxation time of ordered water around macromolecules is substantially larger than that for bulk water. This implies a longer life time for unsolvated electrons and hence a larger probability for inter molecular charge transfers. The basic question remains, is there a physiological significance to structured water?

One of the basic questions in biochemistry is, "How similar are conformations of a molecule in solution and in its crystal?" The hydrophobic phenomenon referred to the fact that certain amino acids tend to be buried inside folded proteins, away from water. This phenomena is thought of as essential to explain protein folding. Indeed, polar amino acid residues tend to be on the outside of protein and non-polar residues inside, in general; although exceptions occur. This phenomena has been explained in terms of accessibility of various residues to the solvent, the proximity of residues to the surface of the protein and the frequency of contact between residues [19–35].

INTERACTION OF WATER WITH DNA

DNA is a polyelectrolyte (a polyanion toward charged species) and a water binding molecule. The hydration of DNA is characterized by two water layers. The first hydration shell is composed of a small number of water molecules (20/nucleotide). This water shell is structurally different than bulk water and ions cannot randomly diffuse in this layer. Five of these water molecules are located at base sites; four, bound at furanose oxygens and phosphodiester linkages; and two, at ionic phosphate sites. The remaining 6 to 7 water molecules form an interface region, hydrogen bonded to water already at site-bonds. The secondary shell has essentially the structure of water found in bulk solvent [36].

The maintenance of DNA conformation is dependent on the absence of perturbing cations and on an intact primary hydration shell. A minimal hydration of 18 moles of water/moles of nucleotide is required to maintain the unperturbed B configuration.

CONCLUSIONS

Many scientists have dedicated their life to the study of water. Yet, a full understanding of this small molecule still eludes us. The presence of hydrogen bonds explains all the properties of water we have uncovered.

REFERENCES

1. Noether, D. L., "Water is . . .," *ChemTech,* 84–90 (February 1982).
2. Morowitz, H., "Ice on the Rocks," *Science, 1982,* 26–27 (September 1982).
3. "Water's Wondrous Talents," *Science Digest, 1982,* 103 (May 1982).
4. Duerden, M. and M. N. Jones, "Amphipathicity an Aggregation in Polypeptie and Protein Systems," *J. Theor. Biol., 75,* 1–19 (1978).
5. Symons, M. C. R., "Water Structure and Reactivity," *Acc. Chem. Res., 14,* 179–187 (1981).
6. *Equilibrium Concepts in Natural Water Systems,* in *Advances in Chemistry Series,* R. F. Gould, ed., American Chemical Society (1967).
7. Luck, A. P., "Water in Biological Systems," in *Inorganic Biochemistry.* Springer-Verlag, Berlin (1974).
8. Angell, C., "Les anomalies de l'eau," *La Recherche, 13* (133), 584–593 (1982).
9. Adams, D. J., "Theory of the Dielectric Constant of Ice," *Nature, 293,* 447–449 (1981).
10. Mayer, E. and P. Bruggeller, "Vitrification of Pure Liquid Water by High Pressure Jet Freezing," *Nature, 298,* 715–718 (1982).
11. Verberne, J. B., H. Loman, J. M. Warman, M. P. deHaas, A. Hummel, and L. Prinsen, "Excess Electrons in Ice," *Nature, 272,* 343–344 (1978).
12. Kolata, G. B., "Water Structure and Ion Binding: A Role in Cell Physiology?" *Science, 192,* 1220–1222 (1976).

13. Bateman, J. G., "Aqueous Artefacts: The Riddle of Bound Water," Office of Naval Research. AD-AO24643 (1976).
14. Aksyonov, S. I., "Regulatory Role of Water in Biological Systems," *Studia Biophysica, 85* (1), 5–6 (1981).
15. Sukhorukov, B. I., "Water in Spatial Organization and Thermal Instability of DNA," *Studia Biophysica, 85* (1), 7–8 (1981).
16. Hodfinger, A. J. *Intermolecular Interactions and Biomolecular Organization.* John Wiley & Sons, New York, NY (1977).
17. Richards, F. M. and T. Richmond, "Solvent, Interfaces and Protein Structure in Molecular Interactions and Activity in Proteins." CIBA Foundation Symposium 60, Excerptia Medica, New York (1978).
18. Tanford, C. *The Hydrophobic Effect: Formation of Micelles and Biological Membranes.* John Wiley & Sons, New York, NY (1973).
19. Wertz, D. H. and H. A. Scheraga, "Influence of Water on Protein Structure. An Analysis of the Preferences of Amino Acid Residues for the Inside or Outside and for Specific Conformations in a Protein Molecule," *Macromolecules, 11* (1), 9–15 (1978).
20. Chothia, C., "Hydrophobic Bonding and Accessible Surface Area in Proteins," *Nature, 248,* 338–339 (1973).
21. Manavalan, P. and P. K. Ponnuswamy, "Hydrophobic Character of Amino Acid Residues in Globular Proteins," *Nature, 275,* 673–674 (1978).
22. Hildebrand, J. H., "Is There a 'Hydrophobic Effect'?" *Proc. Natl. Acad. Sci., 76* (1), 194 (1979).
23. Eisenberg, D. R., M. Weiss, and T. C. Terwilliger, "The Helical Hydrophobic Moment: A Measure of the Amphiphilicity of a Helix," *Nature, 299,* 371–374 (1982).
24. Sinanoglu, O., "The Solvophobic Theory for the Prediction of Molecular Conformations and Biopolymer Bindings in Solutions with Recent Direct Experimental Tests," *Intl. J. of Quantum Chemistry, 18,* 381–392 (1980).
25. Sprang, S., D. Yang, and R. J. Fletterick, "Solvent Accessibility Properties of Complex Proteins," *Nature, 280,* 333–335 (1979).
26. Rackovsky, S. and H. A. Scheraga, "Hydrophobicity, Hydrophilicity, and the Radial and Orientational Distributions of Residues in Native Proteins," *Proc. Natl. Acad. Sci., 74* (12), 5248–5251 (1977).
27. Finney, J. L., "Volume Occupation, Environment and Accessibility in Proteins. The Problem of the Protein Surface," *J. Mol. Biol., 96,* 721–732 (1975).
28. Tanford, C., "Interfacial Free Energy and the Hydrophobic Effect," *Proc. Natl. Acad. Sci., 76* (9), 4175–4176 (1979).
29. Rose, G. D. and S. Roy, "Hydrophobic Basis of Packing in Globular Proteins," *Proc. Natl. Acad. Sci., 77* (8), 4643–4647 (1980).
30. Richards, F. M., "Areas, Volumes, Packing, and Protein Structure," *Ann. Rev. Biophys. Bioeng., 6,* 151–176 (1977).
31. Charton, M. and B. I. Charton, "The Structural Dependence of Amino Acid Hydrophobicity Parameters," *J. Theor. Biol., 99,* 629–644 (1982).
32. Tanford, C., "The Hydrophobic Effect and the Organization of Living Matter," *Science, 200,* 1012–1018 (1978).
33. Hodfinger, A. J., "Solvation Properties of Biopolymers," *Intern. J. Polymeric Mater., 4,* 79–159 (1975).

34. Yamaguchi, T. and C. Nagata, "Molecular Orbital Study on the Effect of the Binding of Water to Base Pair," *J. Theor. Biol., 69,* 693–707 (1977).

35. Hechter, O., "Intracellular Water Structure and Mechanisms of Cellular Transport," *Annals of The New York Academy of Sciences, 125,* 625–646 (1965).

36. Lavery, R. and B. Pullman, "The Electrostatic Field of Oligonucleotides: Relationship to Hydration," *Studia Biophysica, 92* (3), 99–110 (1982).

6

Modified DNA and Proteins

INTRODUCTION

Many proteins of higher organisms are modified with sugar groups. The same proteins produced by microorganisms are typically naked [1]. A great deal of biotechnology involves transferring genes of interest from higher organisms into microbial hosts. The resulting product sometimes lacks the modifications observed in nature. Is this a deadly limitation? Recently, bacteria-produced interferon was shown to retain full activity in the absence of glycosylation. Will this always be the case? It is important for the biotechnologist to understand the role of such protein modifying groups.

Postribosomal modification of proteins is widespread [2] and involves hydroxylation, glycosylation, phosphorylation, methylation, acylation [3,4], cross-link formation [5–7], hydroxylation, ADP-ribosylation, nucleotidylation, carboxylation, etc. [8–12]. DNA and RNA are also modified after their synthesis or duplication [13]. The state of our knowledge in glycosylation, protein and DNA methylation and phosphorylation is reviewed in the following pages.

GLYCOPROTEINS

A large number of proteins have carbohydrates covalently bound to specific peptide portions. Enzymes, hormones, antibody and membrane proteins are glycosated. Glycoproteins are ubiquitous in nature with the exception of bacteria where they have not been observed [14–20].

SUGARS
Glucose
Galactose
Mannose
Fucose
Sialic Acid
N-Acetylgalactosamine
N-Acetylglucosamine

Eleven monosaccharides, most of them hexoses and their derivatives are found attached to proteins. D-glucose, D-galactose, D-mannose, L-fucose, and sialic acids being the most common [15,19]. These sugar moities could potentially store a large amount of information because of their great diversity, conformation features and branching points. More information than can actually be stored in DNA. It is not known if nature looked at that option and discarded it in favor of the universal DNA/RNA blueprint.

The covalent attachment of carbohydrates to proteins involves the C-1 atom of sugar residues and a functional group of an amino acid in the peptide chain. Three types of linkages have been observed: the C-1 of acetylglucosamine and the amide of nitrogen, the O-glycosidic bond between mannose and fucose and serine and threonine respectively; and the linkage between galactose residues and the hydroxyl group of hydroxylysine [15,19].

The addition of sugar moities on specific amino acids is achieved without a template by specific glycosyltransferase enzymes most likely located along the membrane of the endoplasmic reticulum. These enzymes are obviously genetically controlled; hence glycosylation is indirectly under DNA control [15,19].

In most glycoproteins elucided in detail, the sugar chain is preferentially attached to peptide sequences forming β turns [20]. The recognition site for the enzyme attaching N-actylglucosamine to asparagine appears to be Asn—X—Th(Ser); where X is an amino acid.

Non-enzymatic condensation of glucose with proteins to form stable products is a slow but real process under physiological conditions. Although most glucose and protein molecules coexist in tissue without irreversible covalent binding. Non-enzymatic covalent binding is increased in some disease states such as diabetes [15].

The stabilization effect of sugars on protein is not well understood [21–23]. Presumably, the sugars encourage preferential hydration due to their cohesive force increasing the surface tension of the water. In doing so, they prevent the denaturation of the protein.

In most cases, when the sugar moitie is removed from a protein, no change in activity is observed. This finding would indicate a secondary or more subtle role for carbohydrates. Their ubiquity and conservation throughout evolution indicate, nevertheless, a real role of benefit to the organisms.

TYPE OF GLYCOPROTEINS
Enzymes
Hormones
Membrane Components
Serum Proteins
Structural Proteins
Lectins
Immunoglobin
Transport Proteins

Glycoproteins are widespread. They include enzymes (ribonuclease, glucose oxidase, amylase), hormones (thyroglobulin, follicle stimulating hormone, chorionic gonadotrophin), antibodies (I_g—immunoglobins), and membrane proteins (collagen, elastin) [24]. They are part and parcel of plasma proteins (transferrin, ceruloplasmin), the clotting system (prothrombin, fibrinogen), the mucous secretions (mucin, a teeth coating protein) [25,26], the antifreeze property of antarctic fish [27–30] and as cell markers and agglutination factors [31–36].

Carbohydrates, by their sheer presence, may modify several properties of proteins: hydrophobicity, mass, size, electrical charge, mobility, partition between compartments, etc. Most important than all the above is their role as cell markers [15,19,37–39].

The antifreeze proteins are glycoproteins displaying extreme specialization. They are composed of alanine, threonine, n-acetylgalactosamine, and galactose units. The cis-hydroxyl groups of carbon 3 and 4 are responsible for their antifreeze properties. These proteins lower the freezing point of water allowing arctic fish to thrive under what would be otherwise lethal conditions for most organisms [25,28–30].

It is well known that proteins coated with sugar are more resistant to proteolisis and heat denaturation. A few noticeable exceptions bear mentioning. When the sialic acid is removed from ceruloplasmin, it was found to have reduced half-life in circulation. Carbohydrates appear necessary in forming a stable clot (fibrinogen). Follicle stimulating hormone and chorionic gonadotrophin lose their activity when bound sialic acid is removed. Actually in this last case, one finds a direct relationship between CGH activity and sialic acid contents [15,19,40,41]. In spite of this array of interesting features, carbohydrates rarely affect the biological activity of proteins. In most cases, removal of the bound sugar leaves full activity to the protein [42,43].

Carbohydrates as cell markers deserve special attention. It is now recognized that cell surface sugars serve as attachment sites for viruses. Similar molecules (glycoproteins, glycolipids) serve as receptor for lectins. Binding of toxin to surface glycoproteins appears to be a prerequisite to their action. Erythrocytes carry sugar determinants for A, B, and O blood groups as well as for M and N blood group antigens. M and N activity is lost after removal of sialic acid. It has also been suggested that bound sugars could be part of the mysterious cell-cell recognition system so fundamental to life. Cell adhesion is necessary for fertilization, cell differentiation and organogenesis [14,15,19].

A variety of disease states are associated with the absence or accumulation of glycoproteins [14,15,44].

DNA METHYLATION METHYLATED DNA/GENES
Immunoglobulin
α-Globin
Mu MOM
β-Globin
Satellite DNA
Virus Genome
Xenopus Laevis
Chloroplast

The methylation of cytosine is the only post synthetic modification so far observed in the DNA of eukaryotes. The methylation of cytosine occurs predominantly in the dinucleotide 5'CpG3'. The occurrence of 5-mC vary widely: less than 1 mole % in some insects, 2–8 mole % in mammals and in excess of 50 mole % in higher plants. The DNA of prokaryotes contain the modified bases N^6-methyl-adenine and 5-methyl-cytosine [46–48].

Although discovered some 30 years ago and ubiquitous in eukaryotes and prokaryotes; hypothesis explaining the potential role of methylation in DNA control are only now being advanced [48].

The methylation occurs enzymatically after DNA synthesis. It is governed by an enzyme called methylase. The enzyme appears to mostly use methylated DNA strand as a template to methylate the appropriate base in the newly synthesized strand.

THE OBSERVATIONS

A number of observations on methylation have been painstakingly accumulated over the years.

- Mouse satellite DNA is more than twice as methylated than the main DNA. Methyl groups in bovine satellite DNA are clustered [49].
- Specific methylation pattern exist—methylation is not a random phenomena.
- The methylation pattern is symmetric in both DNA strands.
- Methylation is tissue specific [49].
- Methylation patterns are clonally heritable [50].
- A striking inverse correlation has repeatedly been observed between the levels of methylation and the extent these DNA segments are expressed as mRNA [51,52].
- Genes are "undermethylated" in tumor cells [53–55].
- Certain organisms such as SV40 are totally devoid of methyl groups.
- Viral DNA of Herpes Simplex is heavily methylated in latently infected cells but no methyl was found in cells from the productive phase of the infection [56,57].

- Differences in methylation between active and inactive DNA have been documented in Xenopus ribosomal genes [58,59]; viral genes [60]; chicken ovalbumin gene; α- and β-Globin genes of chicken [61,62], human and rabbit; mouse metallothionein-I gene and the J region of the immunoglobin gene [63]; and chloroplasts [64–66].
- Mammalian cells contain a maintenance methylase that insures the retention of the methylation pattern.

THE HYPOTHESES

A number of hypotheses have been formulated to explain the presence of methyl groups and their potential role.

- Gene activity [48,51,52,67,68]
- Cellular differentiation [48,69,70]
- DNA repair [48,71–73]
- Replication blocking [48,74]
- Protection against eukaryotic restriction enzymes [48]

Cellular Differentiation

Today differentiation is thought of in terms of genes being switched on and off (according to some program) and directing the synthesis of specific proteins involved in building the edifice. It has been proposed that selective demethylation (activating gene) is involved in differentiation. Experimental evidences indicate strongly that methylation-demethylation is part of the control mechanism (although not the entire answer) of differentiation [69].

Protection Against Restriction Enzymes

A large number of endonucleases attack specific DNA sequences and cut them open. The restriction enzyme HpaII cleaves the DNA sequence CCGG but fails to cut the methylated sequence C(MeC)GG. On the other hand, another restriction enzyme MspI cleaves the sequence whether it is methylated or not [75].

Replication Blocking

An obvious explanation for the presence of methyl groups would be to block transcription of DNA.

Xenopus laevis chromosomes are all heavily methylated, yet it clearly transcribed methylated sequences. Still transcription could depend on the presence of undermethylated regions. This is supported by the fact that untranscribed satellite DNA regions are generally heavily methylated.

Artificial methylation of gene portion causes some inhibition no matter where the methyl groups were placed (in the promotor or structural part of the gene).

DNA Repair

DNA repair seems to occur only in the unmethylated strand of mismatched DNA duplex structure [71].

Gene Control

The specific pattern of methylation tends to be stably inherited in a cell line. Methylation patterns of sequences not subject to selection pressure are stable over many generations.

A great deal of evidence has been accumulated on the inverse relation between methylation and gene activity. This does not prove itself a cause effect relationship or a sufficient role of methylation in determining gene activity [50,70,76].

Experimental manipulation with in vitro methylation and inhibition of methylation with nucleotide analogue point out to a role for methylation in gene expression. Although some genes with a high level of methylation do not show such a correlation and some genes are completely unmethylated indicating that the control scheme is not universal.

The mechanism involved probably includes modifications in DNA protein interaction.

PHOSPHOLIPIDS METHYLATION

A passing remark should be made about methylated phospholipids. The binding of messengers in the form of neurotransmitter, hormone, lectin, and immunoglobulin to outer surface of cell membranes initiates a series of changes in the membrane. Enzymatic methylation of phospholipids plays an important role in the transduction of signals across membranes in a variety of cells. As phospholipids are methylated, by a phospholipid methyltransferase in the membrane, they are transported from the inside to the outside of the membrane. A second enzyme reversed the process. This flip-flop of methylated phospholipids with associated changes in membrane fluidity facilitate the transmission of signals through membranes [77–85].

PROTEIN METHYLATION

METHYLATED PROTEINS
Histone
Myosin
Actin
Cytochrome C
Opsin
Ribosomal Proteins
Membrane Proteins
Ovalbumine
Flagella Protein
Myofibrillar Protein

The methylation of protein is one of many reversible covalent modifications (others include phosphorylation, adenylation, and proteolytic cleavage) of macromolecules. Methylation has been shown to be involved in a regulatory capacity in bacterial chemotaxis, leukocyte chemotaxis, neuronal function, muscle contraction, electron transport, the synthesis of carotene, and hormone secretion. Bacterial chemotaxis is the best studied of these phenomena [86]. Methylation of basic amino acids by a carboxymethylation (protein carboxyl group) takes place at the ε-amino group of lysine, the guanidinium group of arginine and the imidazole group of histidine. Methylation is achieved reversibly by a methyltransferase enzyme. Methylation of glutamyl and aspartyl residues has also been reported.

Stimuli in the environment are detected by specific receptors on cell surfaces. Methyl-accepting chemotaxis proteins (MCP) are the actual transducer between chemical signals and chemotaxis gene products. Attractants cause the methylation of MCPs to increase; repellents bring about a decrease in methylation. The result is a change in swimming behavior of bacteria when faced with a chemical gradient. A methyesterase subsequently demethylate MCP readying it for the next cycle. The phenomena of dynamic methylation-demethylation as a transducer appears to be a widespread regulatory switch [87–90].

The fungi mucor EF-1α, elongation factor, is the most heavily methylated protein (20% of the tyrosine residues) so far reported in the literature [91,92]. The possibility exists that methylation plays a subtle role in regulating the many interactions between proteins and DNA.

Several other behavioral process regulations seem to be based on methylation. White blood cells use chemotaxis in response to infectious or inflammatory agent. Methylation appears to be the transducer. Methyl-accepting proteins appear to be involved in the secretion of hormones from the pituitary and parathyroid glands.

These preliminary findings indicate a potentially important role for methylation as a transducer for signal transmission and behavioral control at the cellular level [93].

PHOSPHORYLATION OF PROTEIN

PHOSPHORYLATED PROTEINS
Initiation Factor 2a
IgE Receptor
Tyrosine Protein Kinase
Myosin Light Chain
Gastrin-17
Rhodopsin
Insulin Receptor
Glycogen Phosphorylase
Phosphorylase Kinase
Glycogen Synthetase
Histone

Enzyme regulation by reversible phosphorylation appears to be a common mechanism by which intracellular events are made responsible to extracellular signals. Some 35 enzymes and many more proteins are known to be regulated through this pathway. Phosphorylation-dephosphorylation mechanisms have been implicated in the action of myosin in muscle contraction [94–96], in malignant transformation [97–99], in parathyroid secretion [100], in sodium conductance in nerve [101], in acetylcholine receptors action [102,103], in cholesterol mechanism [104, 105], in protein synthesis [106,107], in neural and hormonal control of circulatory activity [108,109], in insulin receptor [110–113], in the action of rhodopsin [114].

Most phosphorylation proteins show multiple site of actions. This provides a method of fine tuning, whereas phosphorylation at one site may amplify or antagonize phosphorylation at another site or modify the rate at which they are phosphorylated. Serine and threonine are the selected amino acid residues involved in phosphorylation allowing to turn proteins on and off.

A few of the most interesting cases of the regulatory role of protein phosphorylation/dephosphorylation cycle include:

Growth

Cell growth, proliferation and differentiation are under the control of a signal. Experimental evidence indicates that this signal may be the phosphorylation of cellular proteins, specifically tyrosine groups of proteins (serine and threonine groups are also targets for phosphorylation) [115,116]. The post translational phosphorylation of specific cellular proteins involved in mediating the action of various hormones and growth factors is well known [106]. The receptors for epidermal growth factor (EGF) and platelet derived growth factors (PDGF) are phosphorylated by associated tyrosine kinase [107]. The phosphorylation of eTF2α results in the inhibition of initiation of protein synthesis in eukaryotic cells. A modest phosphorylation of this factor is sufficient to inhibit initiation [117–119].

Hormonal and Neural Growth

Increasing evidence indicates a major role for enzyme regulation by reversible phosphorylation. Phosphorylase kinase has been identified as the key in both the neural and hormonal mechanism for stimulating glycogenolysis [109].

Protein I is a neuron-specific protein present in presynaptic terminals. Neuro-transmitters increase the state of phosphorylation of the protein and brief periods of nerve conduction cause a large increase in its phosphorylation [108]. Phosphorylated forms of conticotropin (ACTH) have been identified [120]. Proteins extracted from crab nerves exhibit a cycle of phosphorylation/dephosphorylation [101]. The endogenous phsophorylation of electroplax membrane proteins is regulated by calcium and calmodulin [103]. Certain neurotransmitters, serotonine, dopamine, and noradrenaline may exert some of their effects by regulating protein phosphorylation [121,122]. All these indications point out toward a major role for protein phosphorylation in neural and hormonal control.

Malignancy

The most challenging new observation in the field is that five different classes of RNA tumor viruses make proteins that phosphorylate tyrosine. These proteins are essential for transforming a normal cell into a tumor cell [123,125].

Protein kinases that phosphorylate residues were initially found associated with transforming protein of RNA tumor viruses [123]. Several enzymes of cells transformed by viruses contain a higher content of phosphorylated tyrosine residues [124,125].

Cholesterol

The amount of unesterified cholesterol in cells is homeostatically regulated within a very narrow range. One of the most important functions of this cholesterol is in the modulation of phospholipid bilayer fluidity. Experimental studies have shown the role of phosphorylation/dephosphorylation in controlling the three key enzymes involved in cholesterol metabolism. 3-hydroxy 3-methylglutaryl-coenzyme A reductase (the key enzyme involved in cholesterol synthesis) is inhibited by phosphorylation. The two key enzymes involved in cholesterol utilization: acyl-CoA: cholesterol O-acyltransferase and cholesterol 7-hydroxylase are activated by phosphorylation [104,105].

Myosin

The phosphorylation of the myosin light chain protein by a kinase controls the conformation of the molecule which is assembled into filaments. This phosphorylation is necessary to regulate contraction [94–96].

CONCLUSION

It is always amazing to realize how much complexity nature has incorporated in its systems over the long night of evolution. This complexity provides for exquisite fine tuned control and a surprising economy of raw material.

Now that biologists have mastered the DNA code, built gene sequencing and gene synthetizing machines, they encounter a new difficulty in post-transcriptional modifications of proteins. While this difficulty might slow down the mass synthesis of useful proteins for medical and industrial use, it provides an opportunity for additional creativity.

Phosphorylation/dephosphorylation and methylation/demethylation represent on-off switches for the regulatory action of genes and proteins. Sugar moities are cell markers. Cross-links provide stability and protection against degradation. The variety of roles, functions, and mechanisms speak to the flexibility and inventiveness in nature. Modified proteins and DNA segments should not be seen by the practicing biotechnologist as a problem but as an opportunity since these modifications have important roles in stability, chain protection and control strategy. Further, these post-translationally modified structures enrich our knowledge of the methods used by nature and expand the number of action points where the biotechnologist can apply his science and his art.

REFERENCES

1. Lampen, J. O., "Mechanism of Enzyme Secretion by Microorganisms," *Biotechnol. & Bioeng. Symp. No. 3,* 37–41 (1972).

2. Uy, R. and F. Wold, "Postranslational Covalent Modification of Proteins," *Science, 198,* 890–894 (1977).
3. Saxholm, H. J. K., A. Pestana, L. O'Connor, C. A. Sattler, and H. C. Pitot, "Protein Acetylation," *Molecular and Cellular Biochemistry, 46,* 129–153 (1982).
4. Brusilow, S., J. Tinker, and M. L. Batshaw, "Amino Acid Acylation: A Mechanism of Nitrogen Excretion in Inborn Errors of Urea Synthesis," *Science, 207,* 659–661 (1980).
5. Rodakis, G. C. and F. C. Kafatus, "Origin of Evolutionary Novelty in Proteins: How a High-Cysteive Chorion Protein has Evolved," *Proc. Natl. Acad. Sci., 79,* 3551–3555.16 (1982).
6. Ambron, R. T. and L. T. Kremznea, "Post-Translational Modification of Neuronal Proteins: Evidence for Transglutaminase Activity in R2, the Giant Cholinergic Neuron of *Aplysia*," *Proc. Natl. Acad. Sci., 79,* 3442–3446 (1982).
7. Sugaraman, M. and H. Lipke, "Crosslink Precursors from the Dipteran Puparium," *Proc. Natl. Acad. Sci., 79,* 2480–2484 (1982).
8. Richter, C., K. H. Winterholter, S. Baumhuter, H.-R. Lutscher, and B. Moser, "ADP-ribosylation in Inner Membrane of Rat Liver Mitochondria," *Proc. Natl. Acad. Sci., 80,* 3198–3192 (1983).
9. Tokunaga, M., H. Tokunaga, and H. C. Wu, "Post-translational Modification and Processing of *Escherichia coli* Prolipoprotein in vitro," *Proc. Natl. Acad. Sci., 79,* 2205–2259 (1982).
10. Choara, R. K. and V. S. Ananthanarayanan, "Conformational Implications of Enzymatic Proline Hydroxylation in Collagen," *Proc. Natl. Acad. Sci., 79,* 7180–7184 (1982).
11. Thomas, M., J. Janatova, W. R. Gray, and B. F. Tack, "Third Complement of Human Complement: Localization of the Internal Thiolester Bond," *Proc. Natl. Acad. Sci., 9,* 1014–1058 (1982).
12. Bierzynski, A., P. S. Kim, and R. L. Baldwin, "A Salt Bridge Stabilizes the Helix Framed by Isolated C-Peptide of RNase A," *Proc. Natl. Acad. Sci., 79,* 2470–2474 (1982).
13. Horinouchi, S. and B. Weisblum, "Postranscriptional Modification of mRNA Conformation: Mechanism That Regulates Erythromycin-Induced Resistance," *Proc. Natl. Acad. Sci., 77* (12), 7079–7083 (1980).
14. Spiro, R. G., "Glycoproteins: Their Biochemistry, Biology, and Role in Human Disease," *New England Jnl. of Med., 281* (19), 1043-1046 (1969).
15. Spiro, R. G., "Glycoproteins: Their Biochemistry, Biology and Role in Human Disease," *New England Jnl. of Med., 281* (18), 991-1001 (1969).
16. Sharon, N., "Carbohydrates," *Sci. Amer.,* 90–115.
17. Ashwell, G. and A. G. Morell, "The Role of Surface Carbohydrates in the Hepatic Recognition and Transport of Circulating Glycoproteins," *Advents in Enzymology, 41,* 99–128 (1974).
18. Kornfeld, R. and S. Kornfeld, "Comparative Aspects of Glycoprotein Structure," *Ann. Rev. Biochem., 45,* 217–237 (1976).
19. Sharon, N. and H. Lis, "Glycoproteins," *Ann. Rev. Biochem., 41,* 673–702 (1981).
20. Marshall, R. D., "Glycoproteins," *Ann. Rev. Biochem., 41,* 673–702 (1972).
21. Bush, C. A., "Conformational Energy Calculations on Glycosylated Turns in Glycoproteins," *Biopolymers, 21,* 535–545 (1982).

22. Vegarud, G. and T. B. Christensen, "Glycosylation of Proteins: A New Method of Enzyme Stabilization," *Biotechnology and Bioengineering, 17,* 1391-1397 (1975).
23. Timasheff, S. N. and T. Arakawa, "How Sugars Stabilize Proteins Explained," *Biochem., 21,* 6536 (1982).
24. Yamauchi, M., C. Noyes, Y. Kuboki, and G. L. Mechanic, "Collagen Structural Microheterogeneity and a Possible Role for Glycosylated Hydroxylysine in Type I Collagen," *Proc. Natl. Acad. Sci., 79,* 7684–7688 (1982).
25. "Glycoprotein in Saliva May Protect Teeth," *Chemical & Engineering News,* p. 40 (April 11, 1983).
26. Gallagher, J. T. and A. P. Corfield, "Mucin-Type Glycoproteins—New Perspectives On Their Structure and Synthesis," *Trends in Biochemistry, 3,* 38–41 (1978).
27. DeVries, A. L., "Glycoproteins as Biological Antifreeze Agents in Antarctic Fishes," *Science, 172,* 1152–1155 (1971).
28. Feeney, R. E. and D. T. Osuga, "Polar Fish Proteins," *Trends in Biochem. Sci.,* 269–271 (December 1977).
29. Rahmann, H. and R. Hilbig, "Brain Gangliosides Are Involved in the Adaptation of Antarctic Fish to Extreme Low Temperatures," *Naturwissenschaften, 67,* 259–260 (1980).
30. Bunn, H. F. an P. J. Higgins, "Reaction of Monosaccharides with Proteins: Possible Evolutionary Significance," *Science, 213,* 222–224 (1981).
31. Jungery, M., D. Boyle, T. Patel, G. Pasvol, and D. J. Weatherall, "Lectin-Like Polypeptides of P. Falciparum Bind to Red Cell Sialoglycoproteins," *Nature, 301,* 24 (January 1983).
32. Hedo, J. A., M. Kasuga, E. Van Obberghen, J. Roth, and C. R. Kahn, "Direct Demonstration of Glycosylation of Insulin Receptor Subunits by Biosynthetic and External Labeling: Evidence for Heterogeneity," *Proc. Natl. Acad. Sci., 78* (8), 4791–4795 (1981).
33. Rice-Ficht, A. C., K. K. Chen, and J. E. Donelson, "Point Mutations During Generation of Expression-Linked Extra Copy of Trypanosome Surface Glycoprotein Gene," *Nature, 298,* 676–679 (1982).
34. Rosner, M. R., L. S. Grinna, and P. W. Robbins, "Differences in Glycosylation Patterns of Closely Related Murine Leukemia Viruses," *Proc. Natl. Acad. Sci., 77* (1), 67–71 (1980).
35. Vorghese, J. N., W. G. Lauer, and P. M. Colman, "Structure of the Influenza Virus Glycoprotein Antigen Neuraminidase at 2.9A Resolution," *Nature, 305,* 35–40 (1983).
36. Nogueira, N., J. Unkeless, and Z. Cohn, "Specific Glycoprotein Antigens in the Surface of Insect and Mammalian Stages of *Tryponosoma cruzi,*" *Proc. Natl. Acad. Sci., 79,* 1259–1263 (1982).
37. Cardozo, de Almeida, M. Lucia, and M. J. Turner, "The Membrane Form of Variant Surface Glycoproteins of *Trypanosoma brucei,*" *Nature, 302,* 349–352 (1983).
38. Chen, R., "Complete Amino Acid Sequence and Glycosylation Sites of Glycoprotein gp71A of Fried Murine Leukemia Virus," *Proc. Natl. Acad. Sci., 79,* 5788–5792 (1982).
39. Gralnick, H. R., S. B. Williams, and M. E. Rick, "Role of Carbohydrate in Multi-

meric Structure of Factor VIII/Von Willebrand Factor Protein," *Proc. Natl. Acad. Sci., 80,* 2771–2774 (1983).

40. Firestone, G. L., F. Payvar, and K. R. Yamamoto, "Glucocorticoid Regulation of Protein Processing and Compartmentalization," *Nature, 300,* 221–225 (1982).

41. Moyle, W. R., "Role of Carbohydrate of Human Chorionic Gonadotropin in the Mechanism of Hormone Action," *J. Biol. Chem., 250* (23), 9163–9169 (1975).

42. Regoelzi, E., P. A. Chindemi, M. T. Debanne, and P. A. Charlwood, "Partial Resialylation of Human Asiolo Transferrin Type 3 in the Rat," *Proc. Natl. Acad. Sci., 79,* 1116–1120 (1982).

43. Parsons, T. F. and J. G. Pierce, "Oligosaccharide Moieties of Glycoprotein Hormones: Bovine Lutropin Resists Enzymatic Deglycosylation Because of Terminal O-Sulfated N-Acetylhexosamines," *Proc. Natl. Acad. Sci., 77* (12), 7089–7093 (1980).

44. Sandhoff, K., "The Biochemistry of Sphingolipid Storage Diseases," *Angew. Chem. Int. Ed. Engl., 16* (5), 273–285 (1977).

46. Doerfler, W., "DNA Methylation—A Regulatory Signal in Eukaryotic Gene Expression," *J. Gen. Virol., 57,* 1–20 (1981).

47. Ehrlich, M. and R. Y.-H. Wang, "5-Methylcytosine in Eukaryotic DNA," *Science, 212,* 1350–1357 (1981).

48. Razin, A. and A. D. Riggs, "DNA Methylation and Gene Function," *Science, 210,* 604–610 (1980).

49. Sano, H. and R. Sager, "Tissue Specificity and Clustering of Methylated Cytosines in Bovine Satellite I DNA," *Proc. Natl. Acad. Sci., 79,* 3584–3588 (1982).

50. Felsenfeld, G. and J. McGhee, "Methylation and Gene Control," *Nature, 296,* 602–603 (1982).

51. Stein, R., A. Razin, and H. Cedar, "In Vitro Methylation of the Hamster Adenine Phosphoribosyltransferase Gene Inhibits Its Expression in Mouse L Cells," *Proc. Natl. Acad. Sci., 79,* 3418–3422 (1982).

52. Sutter, D. and W. Doerfler, "Methylation of Integrated Adenovirus Type 12 DNA Sequences in Transformed Cells Is Inversely Correlated With Viral Gene Expression," *Proc. Natl. Acad. Sci., 77* (1), 253–256 (1980).

53. "Genes Gone Awry: Clues to Cancer Mechanism Found," *Chemical & Engineering News* (January 17, 1983).

54. Groudine, M., R. Eisenman, and H. Weintraub, "Chromatin Structure of Endogenous Retroviral Genes and Activation by an Inhibitor of DNA Methylation," *Nature, 292,* 311–317 (1981).

55. Feinberg, A. P. and B. Vogelstein, "Hypomethylation Distinguishes Genes of Some Human Cancers From Their Normal Counterparts," *Nature, 301,* 89–90 (1983).

56. Youssoufian, H., S. M. Hammer, S. Hirsch, and C. Mulder, "Methylation of the Viral Genome in an In Vitro Model of Herpes Simplex Virus Latency," *Proc. Natl. Acad. Sci., 79,* 2207–2210 (1982).

57. Simon, D., H. Stuhlmann, D. Jahner, H. Wagner, E. Werner, R. Jaenisch, "Retrovirus Genomes Methylated by Mammalian But Not Bacterial Methylase Are Non-Infectious," *Nature, 304* (21), 275–277 (1983).

58. Maden, B. E. H., "Methylation Map of *Xenopus Laevis* Ribosomal RNA," *Nature, 288,* 293–296 (1980).

59. Harland, R. M., "Inheritance of DNA Methylation in Microinjected Eggs of *Xenopus Laevis*," *Proc. Natl. Acad. Sci., 79,* 2323–2327 (1982).

60. Gupta, K. C., and P. Roy, "Synthesis of Capped and Uncapped Methylated Oligonucleotides by the Virion Transcriptase of Spring Viremia of Carp Virus, a Rhabdovirus," *Proc. Natl. Acad. Sci., 78* (8), 4758–4762 (1981).

61. McGhee, J. D., and G. D. Ginder, "Specific DNA Methylation Sites in the Vicinity of the Chicken B-globin Genes," *Nature, 280,* 419–420 (1979).

62. Haigh, L. S., B. Blanchard Owens, and V. M. Ingram, "DNA Methylation of Remote Sequences in Chicken-globin Gene Expression," *MIT Industrial Liaison Program Report* (June 13, 1983).

63. Mather, E. L. and R. P. Perry, "Methylation Status and DNase I Sensitivity of Immunoglobulin Genes; Changes Associated with Rearrangement," *Proc. Natl. Acad. Sci., 80,* 4689–4693 (1983).

64. Sager, R. and C. Grabowy, "Differential Methylation of Chloroplast DNA Regulates Material Inheritance in a Methylated Mutant of Chlamydomonas," *Proc. Natl. Acad. Sci., 80,* 3025–3029 (1983).

65. Burton, W. G., C. T. Grabowy, and R. Sager, "Role of Methylation in the Modification and Restriction of Chloroplast DNA in Chlamydomonas," *Proc. Natl. Acad. Sci., 76* (3), 1390–1394 (1979).

66. Dyer, T. A., "Methylation of Chloroplast DNA in Chlamydomonas," *Nature, 298,* 422–423 (1982).

67. Gautsch, J. W. and M. C. Wilson, "Delayed de novo Methylation in Teratocarcinoma Suggests Additional Tissue-Specific Mechanisms for Controlling Gene Expression," *Nature, 301,* 32–37 (1983).

68. Gasson, J. C., T. Ryden, and Suzanne Bourgeois, "Role of de novo DNA Methylation in the Glucocorticoid Resistance of a T-Lymphoid Cell Line," *Nature, 302,* 621–623 (1983).

69. Jahner, D., H. Stuhlman,, C. L. Stewart, K. Harbers, J. Lohler, I. Simon, and R. Jaenisch, "De novo Methylation and Expression of retroviral Genomes During Mouse Embryogenesis," *Nature, 198,* 623–628 (1982).

70. Lindahl, T., "DNA Methylation and Control of Gene Expression," *Nature, 290,* 363–364 (1981).

71. Lu, A.-Lien, S. Clark, and P. Modrich, "Methy-Directed Repair of DNA Base-Pair Mismatches *in vitro*," *Proc. Natl. Acad. Sci., 80,* 4639–4643 (1983).

72. Karran, P., T. Hjelmgren, T. Lindahl, "Induction of a DNA Glycosylase for N-Methylated Purines Is Part of the Adaptive Response to Alkylating Agents," *Nature, 296,* 770–773 (1982).

73. Evensen, G. and E. Seeberg, "Adaptation to Alkylation Resistance Involves the Induction of a DNA Glycosylase," *Nature, 296,* 773–775 (1982).

74. Wilson, V. L. and P. A. Jones, "DNA Methylation Decreases in Aging But Not in Immortal Cells," *Science, 220,* 1055–1057 (1983).

75. Topal, M. D., C. A. Hutchinson, III, and M. S. Baker, "DNA Precursors in Chemical Mutagenesis: A Novel Application of DNA Sequencing," *Nature, 298,* 863–865 (1982).

76. "Carcinogens Awaken Sleeping Genes," *New Scientist* (17 March 1983).

77. Hirata, F. and J. Axelrod, "Phospholipid Methylation and Biological Signal Transmission," *Science 209,* 1082–1090 (1980).

78. Hirata, F., S. Toyoshima, J. Axelrod, and M. J. Waxdal, "Phospholipid Methylation: A Biochemical Signal Modulating Lymphocyte Mitogenesis," *Proc. Natl. Acad. Sci., 77* (2), 862–865 (1980).
79. Mato, J. M. and D. Marin-Cao, "Protein and Phospholipid Methylation During Chemotaxis in Dictyostelium Discoideum and Its Relationship to Calcium Movements," *Proc. Natl. Acad. Sci., 76* (12), 6106–6109 (1979).
80. Niwano, M. and B. L. Taylor, "Novel Sensory Adaptation Mechanism in Bacterial Chemotaxis to Oxygen and Phosphotransferase Substrates," *Proc. Natl. Acad. Sci., 79,* 11–15 (1982).
81. Bareis, D. L., V. C. Manganiello, F. Hirata, M. Vaughan, and J. Axelrod, "Bradykinin Stimulates Phospholipid Methylation, Calcium Influx, Prostaglandin Formation and C-AMP Accumulation in Human Fibroblasts," *Proc. Natl. Acad. Sci., 80,* 2514–2518 (1983).
82. Hirata, F. and J. Axelrod, "Phospholipid Methylation and Biological Signal Transmission," *Science, 209,* 1082–1090 (1980).
83. Hirata, F., S. Toyoshima, J. Axelrod, and M. J. Waxdal, "Phospholipid Methylation: A Biochemical Signal Modulating Lymphocyte Mitogenesis," *Proc. Natl. Acad. Sci., 77* (2), 862–865 (1980).
84. Mato, J. M. and D. Marin-Cao, "Protein and Phospholipid Methylation During Chemotaxis in Dictyostelium Discoideum and Its Relationship to Calcium Movements," *Proc. Natl. Acad. Sci., 76* (12), 6106–6109 (1979).
85. Niwano, M. and B. L. Taylor, "Novel Sensory Adapation Mechanism in Bacterial Chemotaxis to Oxygen and Phosphotransferase Substrates," *Proc. Natl. Acad. Sci., 79,* 11–15 (1982).
86. Paik, W. K. and S. Kim *Protein Methylation.* John Wiley and Sons (1980).
87. Segall, J. E., M. D. Manson, and H. C. Berg, "Signal Processing Times in Bacterial Chemotaxis," *Nature, 296,* 856–857 (1982).
88. Kleene, S. J., A. C. Hobson, and J. Adler, "Attractants and Repellents Influence Methylation and Demethylation of Methyl-Accepting Chemotaxis Proteins in an Extract of *Excherichia coli,*" *Proc. Natl. Acad. Sci., 76* (12), 6309–6313 (1979).
89. Toews, M. L., M. F. Goy, M. S. Springer, and J. Adler, "Attractants and Repellents Control Demethylation of Methylated Chemotaxis Proteins in *Escherichia coli,*" *Proc. Natl. Acad. Si., 76* (11), 5544–5548 (1979).
90. Chelsky, D. and F. W. Dahlquist, "Structural Studies of Methyl-Accepting Chemotaxis Proteins of *Escherichia coli*: Evidence for Multiple Methylation Sites," *Proc. Natl. Acad. Sci., 77* (5), 2434–2438 (1980); Sherris, D. and J. S. Parkinson, "Posttranslational Processing of Methyl-Accepting Chemotaxis Proteins in *Escherichia coli,*" *Proc. Natl. Acad. Sci., 78* (10), 6051–6055 (1981).
91. Hiatt, W. R., R. Garcia, W. C. Merrick, and P. S. Sypherd, "Methylation of Elongation Factor 1 From the Fungus Mucor," *Proc. Natl. Acad. Sci., 79,* 3433–3437 (1982).
92. Springer, M. S., M. F. Goy, and J. Adler, "Protein Methylation in Behavioural Control Mechanisms and in Signal Transduction," *Nature, 280,* 279–284 (1979).
93. DeFranco, A. L. and D. E. Koshland, Jr., "Multiple Methylation in Processing of Sensory Signals During Bacterial Chemotaxis," *Proc. Natl. Acad. Sci., 77* (5), 2429–2433 (1980).
94. Craig, R., R. Smith, and J. Kendrick-Jones, "Light-Chain Phosphorylation Con-

trols the Conformation of Vertebrate Non-Muscle and Smooth Muscle Myosin Molecules," *Nature, 302,* 436–439 (1983).

95. Chatrterjee, M. and R. A. Murphy, "Calcium-Dependent Stress Maintenance Without Myosin Phosphorylation in Skinned Smooth Muscle," *Science, 221,* 464–466 (1983).
96. Butler, T. M., M. J. Siegman, S. U. Mooers, and R. J. Barsotti, "Myosin Light Chain Phosphorylation Does Not Modulate Cross-Bridge Cycling Rate in Mouse Skeletal Muscle," *Science, 220,* 1167–1169 (1983).
97. Simon, M. A., T. B. Kornberg, and J. M. Bishop, "Three Loci Related to the src Oncogene an Tyrosine-Specific Protein Kinase Activity in Droisophila," *Nature, 301,* 837–839 (1983).
98. Cooper, J. A., N. A. Reiss, R. J. Schwartz, and T. Hunter, "Three Glycolytic Enzymes Are Phosphorylated at Tyrosine in Cells Transformed by Rous Sarcoma Virus," *Nature, 302,* 218–223 (1983).
99. Fukami, Y. and F. Lipmann, "Reversal of Rous Sarcoma-Specific Immunoglobulin Phosphorylation on Tyrosine (ADP as Phosphate Acceptor) Catalyzed by the src Gene Kinase," *Proc. Natl. Acad. Sci., 80,* 1872–1876 (1983).
100. Bhargava, G., J. Russel, and L. M. Sherwood, "Phosphorylation of Parathyroid Secretory Protein," *Proc. Natl. Acad. Sci., 80,* 878–881 (1983).
101. Schoffeniels, E. and G. Dandrifosse, "Protein Phosphorylation and Sodium Conductance in Nerve Membrane," *Proc. Natl. Acad. Sci., 77* (2), 812–816 (1980).
102. Rephaeli, A. and S. M. Parsons, "Calmodulin Stimulation of $^{45}Ca^{2+}$ Transport and Protein Phosphorylation in Cholinergic Synaptic Vesicles," *Proc. Natl. Acad. Sci., 79,* 5783–5787 (1982).
103. Smilowitz, H., R. A. Hadjian, J. Dwyer, and M. B. Feinstein, "Regulation of Acetylcholine Receptor Phosphorylation by Calcium and Calmodulin," *Proc. Natl. Acad. Sci., 78* (8), 4708–4712 (1981).
104. Scalen, T. J. and A. Sanghvi, "Regulation of Three Key Enzymes in Cholesterol Metabolism by Phosphorylation/Dephosphorylation," *Proc. Natl. Acad. Sci., 80,* 2477–2480 (1983).
105. Gavey, K. L., D. L. Trujillo, and T. Scallen, "Evidence of Phosphorylation/Dephosphorylation of Rat Liver acyl-coA: Cholesterol Acyltransferase," *Proc. Natl. Acad. Sci., 80,* 2171–2174 (1983).
106. Siekierka, J., L. Mauser, and S. Ochoa, "Mechanism of Polypeptide Chain Initiation in Eukaryotes and Its Control by Phosphorylation of the Subunit of Initiation Factor 2," *Proc. Natl. Acad. Sci., 79,* 2537–2540 (1982).
107. Baldwin, G. S., J. Knesel, and J. M. Monckton, "Phosphorylation of Gastrin-17 by Epidermal Growth Factor-Stimulated Tyrosine Kinase," *Nature, 301,* 435–437 (1983).
108. Tsou, K. and P. Greengard, "Regulation of Phosphorylation of Proteins I, III_a, and III_b in Rat Neurohypophysis In Vitro by Electrical Stimulation and by Neuroactive Agents," *Proc. Natl. Acad. Sci., 79,* 6075–6079 (1982).
109. Cohen, P., "The Role of Protein Phosphorylation in Neural and Hormonal Control of Cellular Activity," *Nature, 296,* 613–620 (1982).
110. Kasuga, M., Y. Zick, D. L. Blithe, M. Crettaz, and C. R. Kahn, "Insulin Stimulates Tyrosine Phosphorylation of the Insulin Receptor in a Cell-Free System," *Nature, 298,* 667–669 (1982).

111. Purrello, F., D. B. Burnham, and I. D. Goldfine, "Insulin Receptor Antiserum and Plant Lectinks Mimic the Direct Effects of Insulin on Nuclear Envelope Phosphorylation," *Science, 2321,* 462–464 (1983).

112. Rosen, O. M., R. Herrera, Y. Olowe, L. M. Petruzzelli, and M. H. Cobb, "Phosphorylation Activates the Insulation Receptor Tyrosine Protein Kinase," *Proc. Natl. Acad. Sci., 80,* 3237–3240 (1983).

113. Purrello, F., D. B. Burnham, and I. D. Goldfine, "Insulin Regulation of Protein Phosphorylation in Isolated Rat Liver Nuclear Envelopes: Potential Relationship to mRNA Metabolism," *Proc. Natl. Acad. Sci., 80,* 1189–1193 (1983).

114. Paulsen, R. and J. Bentrop, "Activation of Rhodopsin Phosphorylation Is Triggered by the Lumirhodopsin-Metarhodopsin I Transition," *Nature, 302,* 417–419 (1983).

115. Kolata, G., "Clues to Cell Growth and Differentiation," *Science, 220,* 291 (1983).

116. Kolata, G., "Is Tyrosine the Key to Growth Control?" *Science, 219,* 377–378 (1983).

117. Lerous, A. and I. M. London, "Regulation of Protein Synthesis by Phosphorylation of Eukaryotic Initiation Factor 2α in Intact Reticulocytes and Reticulocyte Lysates," *Proc. Natl. Acad. Sci., 73,* 2147–2151 (1982).

118. Matts, R. L., D. H. Levin, and I. M. London, "Effect of Phosphorylation of the Subunit of Eukaryotic Initiation Factor 2 on the Function of Releasing Factor in the Initiation of Protein Synthesis," *Proc. Natl. Acad. Sci., 80,* 2559–2563 (1983).

119. "Tumor Promoters," *Nature, 302,* 750 (1983).

120. Bennett, H. P. J., C. A. Browne, and S. Solomon, "Biosynthesis of Phosphorylated Forms of Corticotropin-Related Peptides," *Proc. Natl. Acad. Sci., 78* (8), 4713–4717 (1981).

121. Walaas, S. I., D. W. Aswad, and P. Greengard, "A Dopamine- and Cyclic AMP-Regulated Phosphoprotein Enriched in Dopamine-Innervated Brain Regions," *Nature, 301,* 69–71 (1983).

122. Mestikawy, S. El, J. Glowinski, and M. Hamon, "Tyrosine Hydroxylase Activation in Depolarized Dopaminergic Terminals—Involvement of Ca^{2+}-Dependent Phosphorylation," *Nature, 302,* 830–832 (1983).

123. Casnellie, J. E., M. L. Harrison, L. J. Pke, K. E. Hellstron, and E. J. Krebs, "Phosphorylation of Synthetic Peptides by a Tyrosine Protein Kinase From the Particulate Fraction of a Lymphoma Cell Line," *Proc. Natl. Acad. Sci., 76,* 282–286 (1982).

124. Langan, T., "Malignant Transformation and Protein Phosphorylation," *Nature, 286,* 329–330 (1980).

125. Patschinsky, T., T. Hunter, F. S. Esch, J. A. Cooper, and B. M. Sefton, "Analysis of the Sequence of Amino Acids Surrounding Sites of Tyrosine Phosphorylation," *Proc. Natl. Acad. Sci., 79,* 973–977 (1982).

7

Artificial and Synthetic Proteins

INTRODUCTION

N*ature has been* a source of awe and inspiration to the mystic, the poet, the engineer and the chemist. Chemical synthesis is dedicated to creating new molecules of interest to man, based on the diversity afforded by the 96 elements and their many possible combinations. The impetus for creating new chemical species can come from observing nature in action. But nature is conservative. Faced with immense diversity and variety, it concentrates on time-tested strategies and produces much variation on very few themes. Much has already been done in imitating nature. Many drugs and industrial chemicals have been patterned after naturally occurring substances or transient metabolites.

Catalysis is fundamental to all life processes. It is not surprising that chemists are turning to observe nature and enzymes, the prime catalysts, to better understand this phenomena.

More than 6 million chemicals have been recognized by Chemical Abstracts, of which only 63,000 are used in commerce. Thousands of proteins are known today and many more remain to be discovered or elucidated. Why then, artificial and synthetic proteins?

The interest in artificial or synthetic proteins is stimulated by:

1. A need to better understand the catalytic phenomena.
2. The desire to improve on nature in terms of activity, efficiency, cost, etc.
3. The wish to tailor make new catalysts to address burning needs, and
4. The challenge is there.

A number of exercises in developing synthetic proteins, especially enzymes, have taken place recently [1–7]. They can be classified as either coping or modifying naturally occurring enzymes or building catalytic systems that bear little resemblance to enzyme systems. Each approach is described in turn (see Figure 7-1).

ENZYME ACTIVITY

Enzymes, typically accelerate reaction rates by factors of 10^8 to 10^{14} or higher. Non enzymatic catalytic reactions are typically 10^2 to 10^6 times faster than uncatalysed reactions. How do enzymes achieve such a precise specificity and such fabulous catalytic activity? The last question

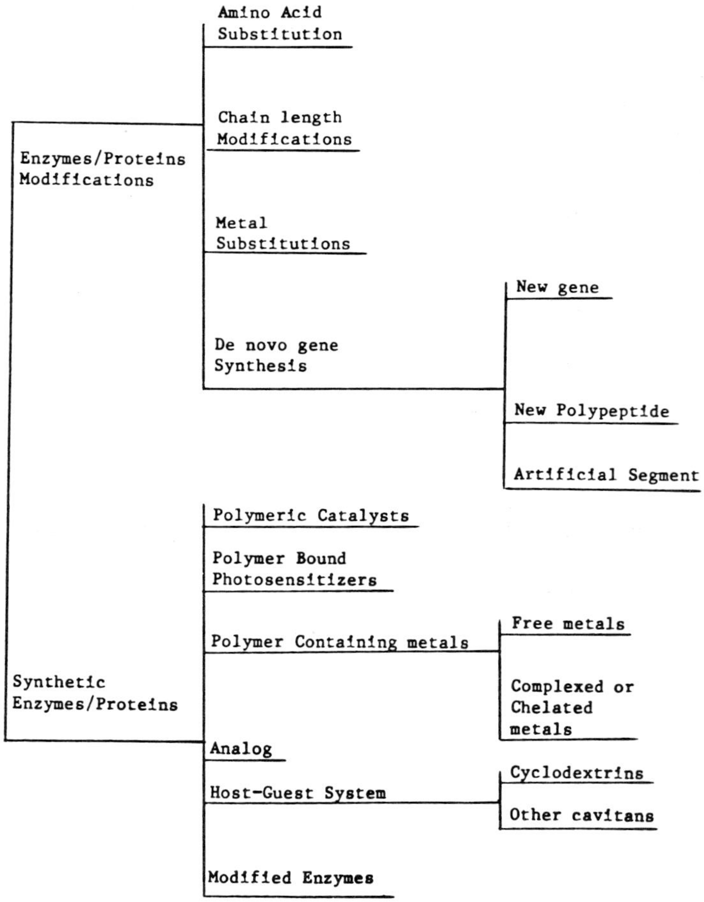

Figure 7-1. Mapping enzymes/proteins.

can be rephrased in terms of enzymes lowering the activation barrier of a given reaction [8–12].

Reaction rate enhancement has been explained by a variety of overlapping and sometimes contradictory mechanisms.

1. The binding between the enzyme and the substrate creates a microenvironment where groups are activated and various geometric strains are incorporated into the complex. The release of such strains by moving to a bound transition state contributes a factor of 10^3–10^4, for instance, to the catalysis of lysozyme. Sheer proximity of enzyme and substrate contributes much to catalysis.
2. The ES binding assures a greater concentration of active form over the inactive conformation. The binding energy observed in ES complex is around -5 Kcal/mol.
3. Enzymes have the ability to stabilize polar states more than bulk water since they can combine the effect of permanent and induced dipoles in a favorable fashion.
4. Desolvation or solvent reorganization plays a substantial role in speeding up reaction. Enzymes can stabilize charges by enveloping an hydrated substrate in an hydrophobic cavity.
5. Entropy losses where two molecules combine to form a tight complex transition state has been estimated to be equivalent to an effective concentration of 10^8M.

The experimental program for the development of an enzyme analog can have a number of approaches. One would be to synthesize compounds which hold the active site groups in almost fixed positions relative to one another. Another approach allows some freedom among the groups to adjust their relative positions. A third approach uses special groups not found at the active site in order to compensate for certain protein effects on such active site properties as electrostatic charge density or ease of electron transfer. These approaches are based on prevailing theories of the detailed mechanisms of enzyme action and include three theories: the template theory (enzyme surface is a three-dimensional negative of the substrate); the unfolding theory (enzyme changes shape on interaction with substrate); and the concerted action theory (several catalytic groups are present at the active site and act together on different portions of the substrate) [13].

ENZYME SPECIFICITY

The exquisite specificity of enzymes for substrates is usually explained by shape complementarity along with dependence on Van der Waal's interactions. Enzymes are not static molecules. Actually enzymatic action always involves a change in nuclear configuration.

PROTEIN MODIFICATIONS

Amino Acid Substitution

For years, scientists have modified the amino acid sequence (a form of artificial mutation) of short peptides and have tested the resulting artificial polypeptide for physiologic activity and pharmaceutic potency and have provided a large harvest of knowledge.

Numerous studies on artificial mutation (amino acid replacement) have been applied to small polypeptides. Such studies have been conducted for somatostatin [14], melanotropin [15], vasopressin [16], and calcitonin [17], to mention only a few. For instance, a variety of modified human and camel β-endorphins were produced by substituting one or more amino acids in the chain of 31 with dramtic impact on potency.

A number of shortened ACTH molecules with strategically (mostly at the NH_2 terminal but with a modified amino acid in position 1) located amino acid substitutions show substantially enhanced activity [18].

A number of omission analogs of polypeptides have also been synthetized and tested for potency [19]. Analogs of β-endorphin show wide variation in analgesic potency, opiate activity, receptor binding activity and immunoreactivity.

Some such substituted polypeptides exist side by side in nature. Met-enkephalin and leu-enkephalin are two such brain opioids, α, β, and γ endorphins are of slightly different length. They all share the first 16 amino acids and differ by the addition of 15 and 1 amino acid respectfully. Their potency is substantially different.

An opposite is vasopressin and oxytocin; two very similar hormones (in terms of structure) but with very dissimilar action: vasopressin constricts arteries and oxytocin promotes the contraction of the uterus [10].

Chain Length Modification

One of the first attempts to create modified enzymes was by shortening polypeptide chains. Fifteen percent of the amino groups of ribonuclease could be removed without loss of activity.

Chain length reduction is not a new area, essentially all secreted proteins exist in a "prepro" form. The pre portion is essentially a signal peptide assisting the protein in crossing membranes. The pro form is generally inactive or of a low activity. It is activated by protelytic cleavage of the pro portion.

Other large proteins like the propiomelanocortin (POMC) are akin to a tool box. The large molecule can be processed into a number of smaller molecules of diverse activities. Different combinations of different peptides, working in cooperation can be released at specific sites (Figure 7-2).

These fragments have diverse activities: β-lipotropin (fat mobilization activity), α-melanotropin (skin pigmentation role), β endorphin (analgesic), ACTH (corticosteroids secretion by adrenal cortex).

Another case is dynorphin which contains [Leu] enkephalin is 300 times more potent than [Leu] enkephalin alone [20]. It is not surprising that many CNS peptides have bifunctional properties: they show hormonal properties and act as possible neurotransmitters [21].

α-melanocyte-stimulating hormone is a linear tridecapeptide that darkens skin reversibly by stimulating pigment granules dispersion within melanophores. A cyclic conformationally restricted analogue has a melanotrophin potency more than 10,000 times that of the native hormone and a prolonged activity relative to the original hormone [22]. By the introduction of covalent conformation and the elimination of amino acid not required, a highly active bicyclic analogue of somatostatin has been built [23].

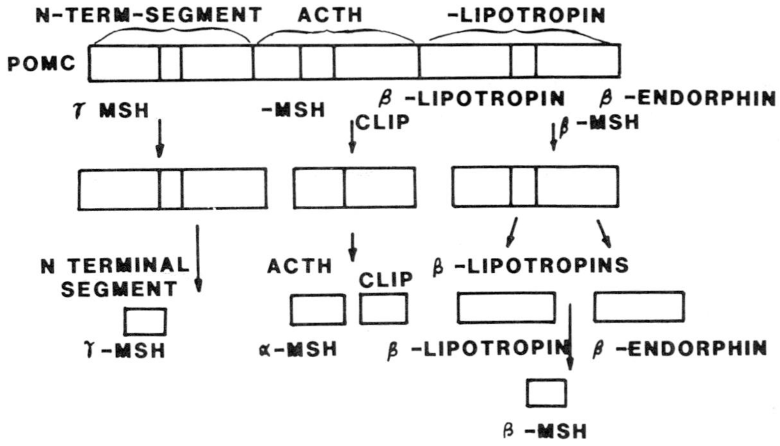

Figure 7-2.

Design of an Analogue Protein

A totally different approach starts with the concept that the action of plasma apolipoprotein A-1 (the principal structural protein of high density lipoprotein (HDL)) is associated with the presence of a small highly amphiphilic α-helical segment [24]. To test the concept, a model docosapeptide with minimum homology to apolipoprotein A-1 was constructed and its ability to activate the lecithin: cholesterol acyltransferase system measured confirming the theory [25,26]. The toxic bee venom melittin is a hexacosapeptide with many properties in common with apolipoprotein A-1. An amphiphilic helix with minimum homology with melittin, while maintaining the hydrophobic-hydrophilic balance was constructed. The dose response curves for erythrocyte lysis were essentially the same for the analogue peptide and the original protein, again confirming the hypothesis [26].

Metals Substitution

A number of native proteins have been modified by exchanging the metals in their active site, creating new semi-synthetic enzymes. Superoxide dismutase is a metallo-protein with Cu and Zn linked by the imidazole group of a histidine residue as the active site. The zinc has been replaced by copper leading to a Cu_2-Cu_2 SOD with similar activity [27]. In other cases important differences in activity are observed [28].

Genes and Proteins Synthesis

The first chemical synthesis of an enzyme was announced in 1965 (synthesis of ribonuclease) [29]. Today we know a better way to synthetize protein—synthetize the gene for that protein.

Genes can now be synthetized by chemical methods in a manual or automated manner. Fragments of 10 to 24 nucleotide long can be obtained this way in a few days to a few weeks of effort. Such genes have been incorporated in living hosts where they reproduce and direct the synthesis of proteins.

What can be done with 20 nucleotide long DNA fragments?

1. Separate fragments can be linked together with the aid of specific enzymes to create a much longer gene.
2. The DNA fragment corresponding to a desired gene is synthetized. Another piece for the opposite strand, but with a short section complementary to the end of the first strand, is synthetized. The double stranded DNA is added to a preparation of DNA polymerase I along

with a sufficient supply of nucleotides. The enzyme sees a piece of DNA in need of repair, and it links together the proper sequence of DNA.
3. Short fragments can serve as probes to identify and localize desired genes.
4. Such fragments can serve as linkage in splicing together desired genes.

A very large number of peptides have been synthetized over the years. The introduction on the market of gene synthetizing machines by Vegas and Biologicals, among others, makes the synthesis of natural and artificial genes a possibility for every biotechnology laboratory.

The list of protein synthetized chemically is rather large: oxytocin, angiotensin, ACTH, gastrin, insulin, secretin, glucacon, ribonuclease, were the earliest attempts.

While it is possible to design a gene to synthetize an enzyme, and rather large polypeptides can be synthetized chemically, other routes to enzyme synthesis are possible.

The primary structure is known for more than 1,000 proteins. The three dimensional structures have been detailed for some 50 proteins. The mode of action has been documented, at the molecular level, for a score of protein/enzyme types.

Strong of this knowledge it should be possible to use existing protein skeleton as the starting point for new enzyme synthesis. Respecting the conserved essential amino acid segments, new segments could be inserted around the active center to give new properties to a natural enzyme.

By specific tinkering with an enzyme structure, we should in theory be able to alter catalytic activity, specificity for substrate, substrate binding strength, micro environment (pH behavior, temperature behavior, etc.), ease of immobilization and alternatively solubility in non-aqueous solvents and susceptibility to inactivation and inhibition.

SYNTHETIC ENZYMES

Enzyme Analogs

Most enzymes are very specific (in terms of reactants and products) catalysts attacking a specific bond, a particular functional group or a stereo isomer. While enzymes are relatively large proteins, the activity center is made up of a limited number of amino acids. Coordinated metals are often part of the catalytic site. Enzymes are difficult and expensive to extract, concentrate and purify. They are fragile and often require costly co-factors for the action. It has also been shown that substantial activity is retained, in a number of cases, after excision of an important part of the peptide chain.

These factors have led chemists to the idea of designing analog enzymes. That is, low molecular weight compound, bearing no relationship to the original enzyme; but performing the same function.

A non-enzymatic RNA polymerase model has been tested. Using preformed copolymer, templates containing all four natural bases direct the non-enzymatic synthesis of their complement starting with a mixture of activated mononucleotide derivatives. Chain length of 8 to 10 nucleotides have been obtained with apparent high fidelity [29].

Analogs of catalase (triethylenetetramine, feraicaron, hydroxic ion complex), tyrosinase (copper-amine-phenol-oxygen complex), pyrocatechase (organic complexing agents coordinated about ferrous iron), and chymotrypsin (imidazole) have shown limited catalytic activity [30,31].

Synthetic peptides have been used as stereo catalysts in organic reactions. The drawbacks are low yield, long reaction time and the large amounts of protein required. The epoxidation of chalcone to the levorotory form has been achieved with optical yields higher than 90 percent with poly [(s)-alanine], poly [5-benzyl-(s)-glutamate], and poly [5-butyl-(s)-glutamate] as catalysts [32].

Optically active polypeptides take part in many reactions such as hydrogenation, oxidation, and reduction [33].

An imidazolyl steroid acts as a catalyst for active esters due to favorable interactions between steroid and ester. When two steroid moities were linked by a p-xylelene bridge, a greater area of hydrophobic contact was achieved with reaction rate enhancement [34].

More recently a number of phosphinomanganese compounds have been synthetized. These complexes bind and release neutral molecular oxygen at ambient temperature and pressure in solid state or in solution and hence mimic the function of hemoglobin and myoglobin.

The complexes (some 200 have been synthetized) have the general formula $MnLX_2$, where L is a phosphine (triethyl phosphine, dimethylphenyl phosphine) and X is an anion (chlorine, bromine, iodine or isothiocyanate) [35,36].

Polymer Containing Metals [37]

About a third of all enzymes contain metals in their active centers. Metals are generally catalysts for chemical reactions; but much less effective than the metal protein complex. Based on these known facts, we can postulate the existence of catalysis when metals are incorporated into organic polymer substrates.

A number of metal ions with coordinate ligands are used in the laboratory to guide and orient the stereochemical course of reactions, achieving the synthesis of chiral compounds [44]. The use of chiral ligand template allows to duplicate the enzymatic lock and key system proposed by Emil Fisher. One of the most dramatic applications of the technique has been the recent synthesis of eight L-hexoses.

By appropriate complexes of transition metals, a variety of enzyme-like reactions have been achieved, e.g., nitrogen fixation, and oxygenase system [38–43].

The interaction between imidazole and transition metal ions is important for the activity of many metalic proteins. For examples, the imidiazolate ion is a bridge between Cu and Zn ions in the enzyme superoxide dismutase; and a histidine bridge has been postulated to link iron heme and copper centers in cytochrome C oxidase [45,46]. Discrete imidazole bridged binuclear metal complexes are important as potential enzyme mimics. Such complexes containing mercury, copper, manganese, iron, cobalt and ruthenium have been synthetized [47]. Lipophilic copper complexes (such as copper (II)-3,5 diisopropylsalicylic acid) have shown to have properties ascribed to superoxide dismutase [48].

The decomposition of hydrogen peroxide to oxygen and water is achieved by a wide variety of reactions but no catalyst competes with catalase in terms of efficiency. A model for the decomposition of hydrogen peroxide is based on chelates formed by TETA (triethylene tetramine) and various metal ions. TETA-Fe^{III} shows remarkable catalytic activity although inferior to the enzyme. The second order rate constant, k_2, for catalyse is 5×10^6 Sec^{-1} per mole of enzyme or 1.3×10^6 Sec^{-1} per mole of hematin iron. The TETA $Fe(OH)_2^+$ system is inferior to catalase by a factor of 10^3 but superior to hemoglobin in efficiency by a factor of about 10^4 [49].

Metals can be incorporated into polymer in the form of gel yielding a readily separatable and usable catalyst. Since the swelling of the gel can be controlled by solvent choice [50], selectivity is enhanced. Microencapsulation of metals in polymers is another avenue that has been pursued.

Finely divided metals have been incorporated and dispersed into solid polymeric matrices. The catalytically active material shows higher selectivity and activity. The solid polymer can be shaped and machined into a variety of forms enhancing its industrial value.

Alternatively, metals have been bound to ion exchange polymers forming complexes or chelates. This is essentially a form of immobilization. Such systems are very complex and the role of diffusion, rigidity of the polymer, number of ligands, porosity, etc., can be untangled with only a great deal of experimental work. These polymer catalysts appear to have valuable properties [51,52].

Polymer Bond Photosensitizers [37]

The oxidation of organic and biological substrates under the influence of light, oxygen and a sensitizer was the path that led to the discovery of the action of a singlet molecular oxygen. Two types of photosensitized oxygenation reactions are observed:

1. The excited sensitizer initiates a free radical-propagated autooxidation.
2. The reactive intermediate is an electronically excited state of molecular oxygen (singlet oxygen) produced by the transfer of energy from the sensitizer to oxygen to produce a reactive metastable state of the oxygen molecule. The term "photosensitized oxidation" is synonymous with oxidations performed with singlet oxygen. The photoexcitation of a sensitizer in the presence of oxygen led to oxidation of a substrate physically removed from the sensitizer. The oxidant is excited molecular oxygen in its metastable singlet state.

$$^1\text{sensitizer} + h\nu \rightarrow {}^3\text{sensitizer}$$

$$^3\text{sensitizer} + {}^3\Sigma_g\text{-}O_2 \rightarrow {}^1\Delta_g O_2 + {}^1\text{sensitizer}$$

$$^1\Delta_g O_2 + \text{acceptor} \rightarrow \text{acceptor} - O_2$$

Rose bengal bound to polystyrene cross linked with divinylbenzene was used for the production of singlets oxygen leading to the oxidation of a number of organic acceptors with high yield.

Among the mononuclear complexes which have been used for the photoproduction of hydrogen from water are a ruthenium bipyridyl $[Ru(bpy)_3]^{+2}$ complex and a zinc porphyrin (Figures 7-3 and 7-4).

The essential reactions of such a system are:

$$S^* + A \rightarrow S^+ + A^-$$

$$S^+ + D \rightarrow S + D^+$$

$$2A^- + 2H^+ \rightarrow 2A + H_2$$

$$D^+ \rightarrow \text{products}$$

where:

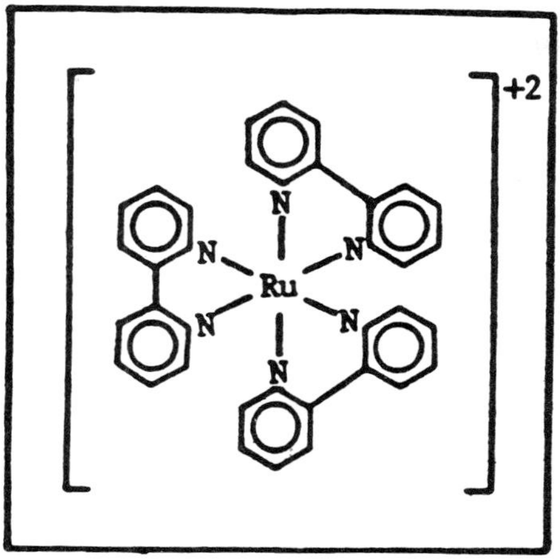

Figure 7-3. Ruthenium 2,2′ bipyridyl complex ion.

S is the photosensitizer and *A* and *D* are electron acceptor and donor, respectively. Harriman [53] and Richoux reported that in the case of the ruthenium bipyridyl sensitizer, with methyl viologen as the electron acceptor, the quantum yield for hydrogen (H_2) approaches 0.15 Grey, et al. [54] have summarized the work of Gretzel and co-workers who have shown that solid ruthenium oxide (RuO_2) catalyzes the evolution of

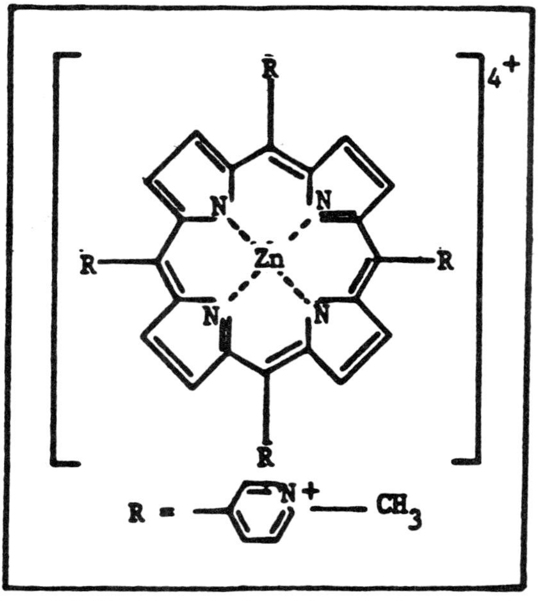

Figure 7-4. Zinc tetramethylpyridyl-porphyrin complex ion.

oxygen in the presence of oxidants such as (RuO_2) catalyzes the evolution of oxygen in the presence of oxidants such as $[Ru(bpy)_3]^{3+}$. $RuyO_2$ combined with $[Ru(bpy)_3]^{2+}$ results in the evolution of both oxygen and hydrogen.

Harriman and Richoux [53] used the zinc porphyrin as the photosensitizer, ethylenediamine teraaceticacid (EDTA) as the irreversible donor, and it is assumed, methyl viologen as the electron acceptor in studies of the photodissociation of water. Platinum, in the form of particles with effective radii of 3–4 mm, on a polyvinyl alcohol (PVA) carrier, was used to catalyze hydrogen evolution in step. The quantum yield for hydrogen (H_2) evolution was 0.3, approximately double the results with $[Ru(bpy)_3]^{+2}$.

Pentaphyrin, an analogue of porphyrins, has been created in the laboratory. Phorphyrins are important molecules. They are the heme of haemoglobin and with some modifications are at the heart of the chlorophyll molecule. Porphyrins are arranged in a cyclic sequence of alternating single and double bond. These π bonds (Kekule's dream) give the molecule its reactivity and stability. Such artificial molecules mimicking the porphyrins will provide considerable help in understanding the mechanism involved and permit building upon nature's start [55].

Cyclodextrins

Enzyme mimics or analogs show many enzyme-like properties. They specifically bind a particular molecule out of a complex milieu; they selectively produce a single product among many possibilities; they accelerate reduction rate by comparison to the similar but uncatalyzed reaction; they release the product to the environment; and remaining unchanged are ready for the next cycle.

Much of the work on artificial enzymes has been conducted with cyclodextrins [56–63]. These are doughnut shaped cavities made up of glucose molecules. α-cyclodextrin (cyclohexaamylose) consists of six glucose units and β-cyclodextrin (cycloheptaamylose) is made up of seven units of glucose. These molecules have cavities of about 5×7 angstroms. They are soluble in water because of the presence of many hydroxyl groups. They have the ability to house small molecules with proper shape and hydrophobic characters.

A number of cyclodextrin-based artificial enzymes have been built and tested. A pyridoxal phosphate—pyridoxamine phosphate coenzyme system as a transaminase mimic showed substrate selectivity and produced optically active amino acids [59,60]; a two imidazole rings attached to a cyclodextrin through a short chain mimic ribonuclease; and a serine protease mimic has been built. These reactions never exceeded 200 fold acceleration in reaction rate.

By a number of architectural and geometric changes such as creating an intrusive floor for the cyclodextrin cavity, or by adding at cap at the rim portions, or by constructing substrate molecules with projecting groups preventing them from binding deeply in the cavity; reaction rate enhancement in the million folds were achieved [62].

Synthetic Greening

Chlorophyll is the basic molecule involved in converting solar energy, CO_2 and H_2O to carbohydrates. The photosynthetic system is admirably complex. The efficiency of photosynthesis is limited to 12% of the light quanta reaching the surface of a green plant. A great deal of interest exists in reengineering the process to achieve higher efficiency [64].

The chlorophyll molecule in all its naked splendor is shown in Figure 7-5. The goal of developing a highly efficient and stable photosynthetic membrane is elusive. Great strides have been made in developing a cell-free system capable of greening (Mg-porphyrin synthesis) *in vitro* at a rate higher than observed *in vivo*. The best such system accumulates chlorophyll at a rate of 93 × 10^3 pmol/100 mg of plastic proteins [65].

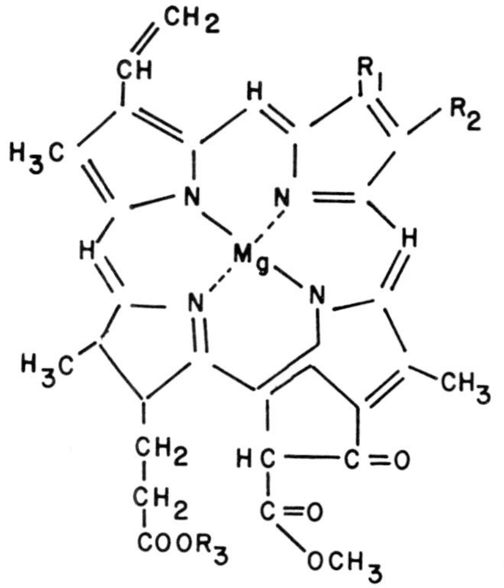

R_1 = - CH_3 R_2 = $-CH=CH_2$ R_3 = FATTY ALCOHOLS
= $-CH_2-CH_3$ = PHYTOL

Figure 7-5. Chlorophylls.

Host-Guest Chemistry

The lessons learned from working with cyclodextrins emphasize the design of new artificial cavities that would bind substrate differently.

A variety of organic compounds with rigid cavities have been synthesized: chorands, cryptans, spherands. These cavities can be egg shape, doughnut shape, cylindrical shape or in shapes approximating, spheres, pots, collars, bowls, vases, saucers, and western saddles. These cavities are available to host larger and larger molecules as potential artificial enzyme complexes [66–68].

Modified Enzymes

A different approach to developing a model catalyst is based on the conversion of simple enzymes which are hydrolytic catalysts into modified enzymes which can catalyze a wide range of important reactions [69]. The process has been called "chemical mutation" [70]. A typical example is the covalent attachment of a redox coenzyme analog at or near the active site of a hydrolytic enzyme creating an easily accessible semi-synthetic oxydo-reductase. Papain has been extensively used as the host enzyme [71]. The idea is to use the binding specificity of the original protein, while using the chemical reactivity of the covalently attached coenzyme. The covalent flavin-papain complex has demonstrated the feasibility of tampering with an enzyme active site to create a new enzyme [72]. Another principle involved here is the well-known fact that coenzymes, in the absence of proteins, often show catalytic activity although usually at a lower rate and with less stereospecificity [73]. A number of analogs of coenzymes have been tested in such a system [74–76].

CONCLUSION

Enzymology is again today a frontier science in many senses. Enzymatic reactions take place on surfaces or at the interface between phases. Enzymology is at the transition point between life and non-life.

REFERENCES

1. Dolphins, D., C. McKenna, Y. Murakami, and I. Tabushi, eds. *Biomimetic Chemistry, Advances in Chemistry Series.* American Chemical Society, Wash., D.C. (1980).

2. Westheimer, F. H., "Models for Enzymatic Systems: An Introduction in Biomimetic Chemistry," D. Dolphins, C. McKenna, Y. Murakami, and I. Tabushi, eds. *Advances in Chemistry Series, 191,* American Chemical Society (1980).
3. Breslow, R., "Adjusting the Lock and Adjusting the Key in Cyclodextrin Chemistry: An Introduction in Biomimetic Chemistry," D. Dolphins, C. McKenna, Y. Murakami, and I. Tabushi, eds. *Advances in Chemistry Series, 191,* American Chemical Society (1980).
4. Fox, J. L., "Synthetic Catalyst Approaches Enzymic Efficiency," *C&EN,* 33–34 (February 14, 1983).
5. "Researchers Close In on Synthetic Enzyme," *C&EN,* 26–27 (April 9, 1979).
6. Pabo, C., "Designing Proteins and Peptides," *Nature,* 301 (1983).
7. Kaiser, E. T. and F. J. Kezdy, "Secondary Structures of Proteins and Peptides in Amphiphilic Environments (A Review)," *Proc. Natl. Acad. Sci., 80,* 1137–1143 (1983).
8. Lipscomb, W., "Intramolecular Interactions, Enzyme Activity and Models in: Molecular Interactions and Activity in Proteins," *Ciba Foundation Symposium 60,* Excertia Medica, NY (1978).
9. Warshel, A., "Energetics of Enzyme Catalysis," *Proc. Natl. Acad. Sci., 75* (11), 5250–5254 (1978).
10. Campbell, I. D., C. M. Dobson, and R. J. P. Williams, "Structure and Energetics of Proteins and Their Active Sites in Molecular Movements and Chemical Reactivity," in *Advances in Chemical Physics,* Vol. 39, R. Lefever and A. Golobeter, eds. John Wiley & Sons, NY (1978).
11. Boger, J. and J. R. Knowles, "To Stabilize a Transition State . . ., in Molecular Interactions and Activity in Proteins," *Excerpta Medica,* Amsterdam (1978).
12. Sinnott, M. L., "Inimitable Enzymes?" *Chem. Br., 15* (6), 293–297 (1979).
13. Rony, P. R., "Synthetic Enzymes," *Biotechnol. & Bioeng. Symp., 3,* 401–412 (1972).
14. Meyers, C. A., D. H. Coy, W. A. Murphy, T. W. Redding, A. Arimura, and D. V. Schally, "[PHe^4] Somatostatin: A Potent, Selective Inhibitor of Growth Hormone Release," *Proc. Natl. Acad. Sci., 77* (1), 577–579 (1980).
15. Schully, A. V., et al., "Isolation, Structure, Biological Characterization and Synthesis of -[Tyr^9] Melanotropin—(9–10) Decapeptide from Pig Hypothalami," *Proc. Natl. Acad. Sci., 77* (7), 3947–3951 (1980).
16. Gitelman, H. S., B. G. Flapper, F. R. Alderman, and W. B. Blythe, "Ala-gly and val-asp-[ARG^8]-vasopressine: Bovine Storage Forms of Arginine Vasopressine with Natriuretic Activity," *Science, 207,* 893–896 (1980).
17. Maier, R., B. Kamber, B. Riniker, and W. Rittel, "Analogues of Human Calcitonin II—Influence of Modifications in Amino Acid Positions 1, 8 and 22 on Hypocalcemic Activity in the Rat," *Hormon. Metab. Res., 7,* 511–514 (1975).
18. Schwyzer, R., "ACTH: A Short Introductory Review," *Ann. NY Acad. Sci., 3,* 26 (1977).
19. Li, C. H., L.-F. Tseng, P. Ferrada, and D. Yamashiro, "-Endorphin: Dissociation of Receptor Binding Activity from Analgesic Potency," *Proc. Natl. Acad. Sci., 77* (4), 2303–2304 (1980).
20. Goldstein, A., S. Tachibana, L. I. Lowney, M. Hunkapiller, and L. Hood, "Dynorphin (1–13), An Extraordinarily Potent Opioid Peptide," *Proc. Natl. Acad. Sci., 76* (12), 6661–6670 (1979).

21. Klee, W. A., "Peptides of the Central Nervous System," in *Advances in Protein Chemistry.* Academic Press, NY (1979).

22. Sawyer, T. K., V. J. Hruby, P. S. Darman, and M. E. Hadley, "[Half-Cys^{4}, Half-Cys^{10}]-α-Melanocyte-Stimulating Hormone: A Cyclic α-Melanotropin Exhibiting Superagonist Biological Activity," *Proc. Natl. Acad. Sci. USA, 79,* 1751–1755 (1982).

23. Veber, D. F., R. M. Freidinger, D. S. Perlow, W. J. Paleveda, Jr., F. W. Holly, R. G. Strachan, R. F. Nutt, B. H. Arison, C. Homnick, W. C. Randall, M. S. Glitzer, R. Saperstein, and R. Hirschmann, "A Potent Cyclic Hexapeptide Analogue of Somatostatin," *Nature, 292,* 55–58 (1981).

24. Yokoyama, S., D. Fukushima, J. P. Kupferberg, F. J. Kezdy, and E. T. Kaiser, "The Mechanism of Activation of Lecithin: Cholesterol Actyltransferase by Apolipoprotein A-I and an Amphiphilic Peptide," *J. Biol. Chem., 255* (15) 7333–7339 (1980).

25. DeGrado, W. F., F. J. Kezdy, and E. T. Kaiser, "Design, Synthesis and Characterization of a Cytotoxic Peptide with Melittin-Like Activity," *J. Amer. Chem. Soc., 103,* 679–681 (1981).

26. Kupferberg, J. P., D. J. Kroon, S. Yokoyama, D. Fukushima, B. W. Shen, and E. T. Kaiser, "Synthetic Apolipoprotein Model Peptides: Structural Determinants of Peptide-Phospholipid Interactions," *Pepti Structure Biol. Func. Proc. Am. Symposium (6th)* (1979).

27. Ibers, J. A. and R. H. Holm, "Modeling Coordination Sites in Metallobiomolecules," *Science, 209* (4453) 223–235 (1980).

28. Fee. J. A. and R. G. Briggs, "Studies on the Reconstruction of Bovine Erythrocyte-superoxide Dismutase. V. Preparation and Properties of Derivatives in which Both Zinc and Copper Sites Contain Copper," *Biochimica et Biophysical Acta, 400,* 439–450 (1975).

29. Inoue, T. and L. E. Orgel, "A Nonenzymatic RNA Polymerase Model," *Science, 219,* 859–862 (1983).

30. Welch, R. C. W. and H. F. Rase, "Selectivity Characteristics of a Geometrically Designed Heterogeneous Catalyst," *I&EC Fundamentals,* 389–397 (1968).

31. Wingard, L. B., Jr. and R. K. Finn, *Chemical Engineering Progress Symposium Series, 62* (69), 30–36 (1966).

32. Julia, S., J. Masana, and J. C. Vega, " 'Synthetic Enzymes' Highly Stereoselective Epoxidation of Chalcone in a Triphasic Toluene-Water-Poly[(S)-alanine] System," *Chem., 92,* 969 (1980).

33. Julia, S., J. Guixer, J. Masana, J. Rocas, S. Colonna, R. Annuziata, and H. Molinari, "Synthetic Enzymes. Part 2. Catalytic Asymmetric Epoxidation by Means of Polyamino-acids in a Triphase System," *J. Chem. Soc. Perkin Trans., I,* 1317–1324 (1982).

34. Guthrie, J. P., P. A. Cullimore, R. S. McDonald, and S. O'Leary, "Large Hydrophobic Interactions with Clearly Defined Geometry. A Dimeric Steroid with Catalytic Properties," *Can. J. Chem., 60* (6), 747–764 (1982).

35. O'Sullivan, D., "Manganese (II) Complexes Mimic Function of Heme," *C&EN, 4,* 24–25 (Dec. 1978).

36. "One Step Nearer to Synthetic Haemoglobin," *New Scientist, 14,* 75 (April 1983).

37. Muetterties, E. L. and M. J. Krause, "Catalysis by Molecular Metal Clusters," *Angew. Chem. Int. Ed. Engl., 22,* 135–148 (1983).

38. Westheimer, F. H., "Models for Enzymic Systems," in *Biomimetic Chemistry,* D.

Dolphin, C. McKenna, Y. Murakami, and Iwao Tabushi, eds., American Chemical Society (1980).

39. James, B. R., "Activation of Dioxygen Using Group VIII Metal Complexes," in *Biomimetic Chemistry,* D. Dolphin, C. McKenna, Y. Murakami, and I. Tabushi, eds., American Chemical Society (1980).
40. Tabushi, I. and N. Koga, "Oxygen Activation: Participation of Metalloenzymes and Related Metal Complexes," in *Biomimetic Chemistry,* D. Dolphin, C. McKenna, Y. Murakami, and I. Tabushi, eds., American Chemical Society (1980).
41. Yoshida, Z.-I., H. Sugimoto, and H. Ogoshi, "Oxygen Activation in Oxygenase Systems," in *Biomimetic Chemistry,* D. Dolphin, C. McKenna, Y. Murakami, and I. Tabushi, eds., American Chemical Society (1980).
42. Newton, W. E., "Chemical Approaches to Nitrogen Fixation," in *Biomimetic Chemistry,* D. Dolphin, C. McKenna, Y. Murakami, and I. Tabushi, eds., American Chemical Society (1980).
43. Chatt, J., "A Possible Mimic of the Nitrogenase Reaction," in *Biomimetic Chemistry,* D. Dolphin, C. McKenna, Y. Murakami, and I. Tabushi, eds., American Chemical Society (1980).
44. Maugh, T. H., II, "Catalysts That Break Nature's Monopoly," *Science, 221,* 351–354 (1983).
45. Haddad, M. S. and D. N. Hendrickson, "Magnetic Exchange Interactions in Binuclear Transition-Metal Complexes. 15. Copper (II) and Nickel (II) Complexes Bridged by Imidazolate, Benzimidazolate, Biimidazolate, and Bibenzimidazolate Ions," *Inorganic Chemistry, 17* (9), 2622–2630 (1978).
46. O'Young, C.-L., J. C. Dewan, H. R. Lilienthal, and S. J. Lippard, "Electron Spin Resonance, Magnetic, and X-Ray Crystallographic Studies of a Binuclear, Imiodazolate Bridged Copper(II) Complex, $[(TMDT)_2Cu_2(im)(ClO_4)_2](ClO_4)$," *J. Amer. Chem. Soc., 100* (23), 7291–7300 (1978).
47. Coughlin, P. K., J. C. Dewan, S. J. Lippard, E. Watanabe, and J. Lehn, "Synthesis and Structure of the Imidazolate Bridged Dicopper(II) Ion Incorporated into a Circular Cryptate Macrocycle," *J. Amer. Chem. Soc., 101* (1), 265–266 (1979).
48. Kensler, T. W., D. M. Bush, and W. J. Kozumbo, "Inhibition of Tumor Promotion by a Biomimetic Superoxide Dismutase," *Science, 221,* 75–77 (1983).
49. Wang, J. H., "On the Detailed Mechanism of a New Type of Catalase-Like Action," *J. Am. Chem. Soc., 77,* 4715–4719 (1955).
50. Manecke, G. and W. Storck, "Polymeric Catalysts," *Angew. Chem. Int. Ed. Engl., 17,* 657–670 (1978).
51. Maugh, T. H., II, "Clusters Provide Unusual Chemistry," *Science, 220,* 592–595 (1983).
52. "Metal Catalysts Snared in Synthetic Vesicles," *C&EN,* 57 (April 18, 1983).
53. Gray, H. B. and A. W. Maverick, *Science, 214,* 1201 (1981).
54. Harriman, A. and M.-C. Richaux, *Journal of Photochemistry, 15,* 335–336 (1981); "Mimicking Photosynthesis in the Laboratory," *New Scientist, 17,* 439 (February 1983).
55. Milgrom, L., "Aromatics on the Edge of Stability," *New Scientist, 16,* 779 (June 1983).
56. Cram, D. J. and H. E. Katz, "An Incremental Approach to Hosts That Mimic Serine Proteases," *Jnl. Amer. Chem. Soc., 105,* 135–137 (1983).

57. Breslow, R., "Artificial Enzymes," *Science, 218,* 532–537 (1982).
58. Hingerty, B. and W. Saenger, "Disorder in a Hydrophobic Cage Illustrated by X-Ray Structure of -Cyclodextrin Methanol Pentahydrate Adduct," *Nature, 255,* 396–397 (1975).
59. Breslow, R. M. and Hammond, M., "Selective Transamination and Optical Induction by a -Cyclodextrin- Pyridoxamine Artivicial Enzyme," *J. Am. Chem. Soc., 102* (1), 421–422 (1980).
60. Breslow, R. and A. W. Czarnik, "Transaminations by Pyridoxamine Selectively Attached at C-3 in -Cyclodextrin," *J. Am. Chem. Soc., 105,* 1390–1391 (1983).
61. VanEtten, R. L., J. F. Sebastian, G. A. Clowes, and M. L. Bender, "Acceleration of Phenyl Ester Cleavage by Cycloamyloses. A Model for Enzymatic Specificity," *J. Amer. Chem Soc., 89* (13), 3242–3253 (1967).
62. Tabushi, I., "Functionalized Cyclodextrins as Artificial Enzymes," *Stud. Org. Chem. (Amsterdam), 10,* 275–280 (1982).
63. Tabushi, I., K. Shimokawa, N. Shimizu, H. Shirakata, and K. Fujita, "Caped Cyclodestrin," *J. Am. Chem. Soc., 98,* 7855–7856 (1976).
64. "Mimicking Photosynthesis in the Laboratory," *New Scientist, 17,* 439 (February 1983).
65. Rebeiz, C. A., H. Daniell, and J. R. Mattheis, "Chloroplast Bioengineering: The Greening of Chloroplast in vitro," *Biotechnology and Bioengineering Symp. No. 12,* 413–439 (1982).
66. Cram, D. J. and J. M. Cram, "Host-Guest Chemistry," *Science, 183* (4127) 803–809 (1974).
67. Cram, D. J., "Cavitands: Organic Hosts with Enforced Cavities," *Science, 219* (4589) 1177–1183 (1983).
68. Cram, D. J., I. B. Dicker, C. B. Knobler, and K. N. Trueblood, "Spherand Hosts Containing Cyclic Urea Units," *Jnl. Amer. Chem. Soc., 104,* 6830–6831 (1982).
69. Levine, H. L. and E. T. Kaiser, "Oxidation of Dihydronicotinamines by Flavopapain," *J. Amer. Chem. Soc., 100* (24), 7670–7677 (1978).
70. Slama, J. T., S. R. Oruganti, and E. T. Kaiser, "Semisynthetic Enzymes: Synthesis of a New Flavopapain with High Catalytic Efficiency," *J. Am. Chem. Soc., 103,* 6211–6213 (1981).
71. Levine, H. L., Y. Nakagawa, and E. T. Kaiser, "Flavopapain: Synthesis and Properties of Semi-synthetic Enzymes," *Biochemical and Biophysical Research Com., 76* (1), 64–70 (1977).
72. Kaiser, E. T., H. L. Levine, T. Otuski, H. E. Fried, and R.-M. Dupeyre, "Studies on the Mechanism of Action and Stereochemical Behavior of Semisynthetic Model Enzymes," *Adv. Chem. Ser., 191,* 35–48 (1980).
73. Dolphin, D., "Model Studies and the Biochemical Function of Coenzymes," in *Biomimetic Chemistry,* D. Dolphin, C. McKenna, Y. Murakami, and I. Tabushi, eds., American Chemical Society (1980).
74. Toraya, T. and S. Fukui, "Structure-Function Relationship of Vitamin B_{12} Coenzyme (Adenosylcobalamin) in the Diol-Dehydrase System," in *Biomimetic Chemistry,* D. Dolphin, C. McKenna, Y. Murakami, and I. Tabushi, eds., American Chemical Society (1980).
75. Kaiser, E. T., H. L. Levine, T. Otuski, H. E. Fried, and R.-M. Dupeyre, "Studies on

the Mechanism of Action and Stereochemical Behavior of Semisynthetic Model Enzymes," in *Biomimetic Chemistry,* D. Dolphin, C. McKenna, Y. Murakami, and I. Tabushi, eds., American Chemical Society (1980).

76. Chao, Y., G. R. Weisman, G. D. Y. Sogah, and D. J. Cram, "Host-Guest Complexation. 21. Catalysis and Chiral Recognition Through Designed Complexation of Transition States in Transacylations of Amino Ester Salts," *J. Amer. Chem. Soc., 101* (7), 4948–4958 (1979).

8

Enzymes

INTRODUCTION

Enzymes are marvelously engineered catalysts. They are ubiquitous in nature; and they accelerate, under mild conditions, every chemical reaction known to man. In spite of these and other attributes, their use in American industry is very limited.

The purpose of this chapter is to review our knowledge of enzymes; to catalogue their major uses and applications; and to look into the near future.

ENZYMES CLASSIFICATION

The commission on Enzymes of the International Union of Biochemistry has devised a classification system for enzymes. Enzymes are placed in either of six broad classes.

1. Oxidoreductase
2. Transferases
3. Hydrolases
4. Lyases
5. Isomerases
6. Ligases (synthetases)

Each class is further divided into subclasses based on the reaction involved, the need for coenzymes and cofactors, the nature of the isomerization and the types of bonds involved.

Table 8-1 is a classification of the enzymes. This table will serve as a

TABLE 8-1. Enzyme Classification.

Classification	Examples
1. Oxidoreductases	
1.1 Acting on the CH-OH group of donors	
1.1.1 with NAD or NADP as acceptor	
	1.1.1.1 Alcohol dehydrogenase
	1.1.1.27 L-lactate dehydrogenase
1.1.2 with a cytochrome as an acceptor	
1.1.3 with O_2 as acceptor	
	1.1.3.4 glucose oxidase
	1.1.3.9 galactose oxidase
1.1.99 with other acceptors	1.10.3.2 laccase
	1.11.1.6 catalase
	1.11.1.7 peroxidase
	1.13.11.12 lipoxidase
1.2 Acting on the aldehyde or keto-group of donors	
1.2.1 with NAD or NADP as acceptor	
	1.2.32 xanthine oxidase
1.2.2 with a cytochrome as an acceptor	
1.2.3 with O_2 as acceptor	
1.2.4 with lipoate as acceptor	
1.2.99 with other acceptors	
1.3 Acting on the CH-CH group of donors	
1.3.1 with NAD or NADP as acceptor	
1.3.2 with a cytochrome as an acceptor	
1.3.3 with O_2 as acceptor	
1.3.99 with other acceptors	
1.4 Acting on the $CH\text{-}NH_2$ group of donors	
1.4.1 with NAD or NADP as acceptor	
1.4.3 with O_2 as acceptor	
	1.4.3.2 L-amino acidoxidase
1.5 Acting on the C-NH group of donors	
1.5.1 with NAD or NADP as acceptor	
1.5.3 with O_2 as acceptor	
1.6 Acting on reduced NAD or NADP as donor	
1.6.1 with NAD or NADP as acceptor	
1.6.2 with a cytochrome as an acceptor	
1.6.4 with a disulfide compound as acceptor	
1.6.5 with a quinone or related compound as acceptor	
1.6.6 with a nitrogenous group as acceptor	
1.6.99 with other acceptors	
1.7 Acting on other nitrogenous compounds as donors	

(continued)

TABLE 8-1. (continued).

Classification	Examples
1.7.3 with O_2 as acceptor	
	1.7.3.3 Uricase
1.7.99 with other acceptors	
1.8 Acting on sulfur groups of donors	
1.8.1 with NAD or NADP as acceptor	
1.8.3 with O_2 as acceptor	
1.8.4 with a disulfide compound as acceptor	
1.8.5 with a quinone or related compound as acceptor	
1.8.6 with a nitrogenous group as acceptor	
1.9 Acting on heme groups of donors	
1.9.3 with O_2 as acceptor	
	1.9.3.1 Cytochrome c oxidase
1.9.6 with a nitrogenous group as acceptor	
1.10 Acting on diphenols and related substances as donors	
1.10.3 with O_2 as acceptor	
1.11 Acting on H_2O_2 as acceptor	
1.12 Acting on hydrogen as donor	
1.13 Acting on single donors with incorporation of oxygen (oxygenases)	
1.14 Acting on paired donors with incorporation of oxygen into one donor (hydroxylases)	
1.14.1 using reduced NAD or NADP as one donor	
1.14.2 using ascorbate as one donor	
1.14.3 using reduced pteridine as one donor	
2. Transferases	
2.1 Transferring one-carbon groups	
2.1.1 methyltransferases	
2.1.2 hydroxymethyl-, formyl-, and related transferases	
2.1.3 carboxyl- and carbamyltransferases	
2.1.4 amidinotransferases	
2.2 Transferring aldehydic or ketonic residues	
2.3 Acyltransferases	
2.3.1 acyltransferases	
2.3.2 aminoacyltransferases	
2.4 Glycosyltransferases	
2.4.1 hexosyltransferases	
	2.4.1.15 dextran sucrase

(continued)

TABLE 8-1. (continued).

Classification	Examples
2.4.2 pentosyltransferases	
2.5 Transferring alkyl or related groups	
2.6 Transferring nitrogenous groups	
2.6.1 aminotransferases	
2.6.3 oximinotransferases	
2.7 Transferring phosphorus-containing groups	
2.7.1 phosphotransferases with an alcohol group as acceptor	
	2.7.1.1 Hexokinase
	2.7.1.40 Puryvate kinase
2.7.2 phosphotransferases with a carboxyl group as acceptor	
2.7.3 phosphotransferases with a nitrogenous group as acceptor	
	2.7.3.2 creatine kinase
2.7.4 phosphotransferases with a phospho group as acceptor	
2.7.5 phosphotransferases, apparently intramolecular	
2.7.6 pyrophosphotransferases	
2.7.7 nucleotidyltransferases	
	2.7.7.7 DNA polymerase
2.7.8 transferases for other substituted phospho groups	
2.8 Transferring sulfur-containing groups	
2.8.1 sulfurtransferases	
2.8.2 sulfotransferases	
	2.8.2.10 luciferase
2.8.3 CoA-transferases	
3. Hydrolases	
3.1 Acting on ester bonds	
3.1.1 carboxylic ester hydrolases	
	3.1.1.3 lipase
	3.1.1.7 actylcholinesterase
3.1.2 thiolester hydrolases	
3.1.3 phosphoric monoester hydrolases	
	3.1.3.1 alkaline phosphatase
3.1.4 phosphoric diester hydrolases	
	3.1.4.22 ribonuclease
3.1.5 triphosphoric monoester hydrolases	
3.1.6 sulfuric ester hydrolases	
	3.1.26.1 RNAase
3.2 Acting on glycosyl compounds	
3.2.1 glycoside hydrolases	

(continued)

TABLE 8-1. (continued).

Classification	Examples
	3.2.1.1 α-amylase
	3.2.1.2 β-amylase (diastase)
	3.2.1.3 glucoamylase
	3.2.1.4 cellulase
	3.2.1.20 maltase
	3.2.1.21 β-glucosidase
	3.2.1.22 α-galactosidase
	3.2.1.23 lactase (β-galactosidase)
	3.2.1.26 invertase
3.2.2 hydrolyzing N-glycosyl compounds	
3.2.3 hydrolyzing S-glycosyl compounds	
3.3 Acting on ether bonds	
3.3.1 thioether hydrolases	
3.4 Acting on peptide bonds (peptide hydrolases)	
3.4.1 α-amino-acyl-peptide hydrolases	
3.4.2 peptidyl-amino-acid hydrolases	
	3.4.2.5 carboxypeptidase
3.4.3 dipeptide hydrolases	
3.4.4 peptidyl-peptide hydrolases	
	3.4.11.8 peptidase
	3.4.21.1 chymotrypsin
	3.4.21.4 trypsin
	3.4.21.7 streptokinase
	3.4.21.14 subtilisin
	3.4.22.2 papain
	3.4.22.3 ficin
	3.4.22.4 bromelain
	3.4.23.1 pepsin
	3.4.23.4 rennin
	3.4.24.4 protease
3.5 Acting on C-N bonds other than peptide bonds	
3.5.1 in linear amides	
	3.5.1.1 asparaginase
	3.5.1.5 urease
3.5.2 in cyclic amides	
	3.5.1.11 penicillin acylase
	3.5.2.6 penicillinase
3.5.3 in linear amidines	
	3.5.3.1 arginase
3.5.4 in cyclic amidines	
3.5.5 in cyanides	
3.5.99 in other compounds	
3.6 Acting on acid anhydride bonds	

(continued)

TABLE 8-1. (continued).

Classification	Examples
3.6.1 in phosphoryl-containing anhydrides	
	3.6.1.13 ATPase
3.7 Acting on C-C bonds	
3.7.1 in ketonic substances	
3.8 Acting on halide bonds	
3.8.1 in C-halide compounds	
3.8.2 in P-halide compounds	
3.9 Acting on P-N bonds	
4. Lyases	
4.1 Carbon-carbon lyases	
4.1.1 carboxy layses	
	4.1.1.1 carbonyl anhydrase
4.1.2 aldehyde layses	
	4.1.2.13 aldolase
4.1.3 ketoacid lyases	
	4.1.99.2 β-tyrosinase
4.2 Carbon-oxygen lyases	
4.2.1 hydro lyases	
	4.2.1.2 fumarase
	4.2.2.1 hyalurodinase
4.2.99 other carbon-oxygen lyases	
4.3 Carbon-nitrogen lyases	
4.3.1 ammonia lyases	
	4.3.1.1 aspartase
4.3.2 amidine lyases	
4.4 Carbon-sulfur lyases	
4.5 Carbon-halide lyases	
4.99 Other lyases	
5. Isomerases	
5.1 Racemases and epimerases	
5.1.1 acting on amino acids and derivatives	
5.1.2 acting on hydroxy acids and derivatives	
5.1.3 acting on carbohydrates and derivatives	
5.1.99 acting on other compounds	
5.2 Cis-trans isomerases	
5.3 Intramolecular oxidoreductases	
5.3.1 interconverting aldoses and ketoses	
	5.3.1.5 glucose isomerase
5.3.2 interconverting keto and enol groups	
5.3.3 transposing C-C bonds	

(continued)

TABLE 8-1. (continued).

Classification	Examples
5.4 Intramolecular transferases	
5.4.1 transferring acyl groups	
5.4.2 transferring phosphoryl groups	
5.4.99 transferring other groups	
5.5 Intramolecular lyases	
5.99 Other isomerases	
6. Ligases	
6.1 Forming C-O bonds	
6.1.1 amino-acid-RNA ligases	
6.2 Forming C-S bonds	
6.2.1 acid-thiol ligases	
6.3 Forming C-N bonds	
6.3.1 acid-ammonia ligases (amide synthetases)	
6.3.2 acid-amino-acid ligases (peptide synthetases)	
6.3.3 cyclo-ligases	
6.3.4 other C-N ligases	
6.3.5 C-N ligases with glutamine as N-donor	
6.4 Forming C-C bonds	

Source: Modified from 1,2.

framework for our discussion. The most common enzymes, repeatedly mentioned in the text, have been superimposed unto the classification scheme [1,2].

ENZYME ACTIVITY

Enzymes perform a wide range of chemical modifications and rearrangements as shown in the examples below.

$$\text{Ethanol} \xrightarrow[\text{Dehydrogenase}]{\text{Alcohol}} \text{Acetaldehyde}$$

$$H_2O_2 \xrightarrow[\text{catalase}]{} O_2 + H_2O$$

$$\text{Xanthine} + H_2O + O_2 \xrightarrow[\text{Oxidase}]{\text{Xanthine}} \text{Urate} + H_2O$$

$$\text{Glucose} + O_2 + H_2O \xrightarrow[\text{oxidase}]{\text{glucose}} H_2O_2 + \text{Gluconic acid}$$

$$\text{Starch} \xrightarrow[\alpha\text{-amylase}]{} \text{glucose} + \text{maltose}$$

$$\text{Starch} \xrightarrow[\beta\text{-amylase}]{} \text{maltose} + \text{dextrin}$$

$$\text{Sucrose} \xrightarrow[\text{invertase}]{} \text{glucose} + \text{fructose}$$

$$\text{Lactose} \xrightarrow[\text{lactase}]{} \text{glucose} + \text{fructose}$$

$$\text{Casein} \xrightarrow[\text{rennin}]{} \text{paracasein}$$

Enzymes are characterized by the speed of the reaction they catalyzed and their sensitivity to environmental conditions.

Enzyme Kinetics

Enzymes are catalysts speeding up chemical reactions. The reaction rate of an enzymatic reaction tends toward a maximum value V_{max} characteristic of the enzyme system.

In the simplest of all cases: One substrate (S), one enzyme (E), one product (P); the Michaelis-Menten formulation is the following:

$$E + S \underset{k_{-1}}{\leftrightarrow} (ES) \overset{k_2}{\rightarrow} E + P$$

Under steady state condition

$$K_1 [E][S] = k_{-1} k_2 [ES]$$

$$\text{or } [ES] = \frac{[E][S]}{K_m}$$

where K_m, the Michaelis constant is given by

$$K_m = (K_{-1} + K_2)/k_1$$

The total quantity of enzymes present is $[E]_t = [E] + [ES]$. Substituting above, yield:

$$[ES] = [E]_t \frac{[S]}{[S] + K_m}$$

The reaction rate V of the product formed is dependent on the concentration of the enzyme substrate complex

$$V = k_2[ES]$$

$$V = k_2[E]_t \frac{[S]}{[S] + K_m}$$

The maximal reaction rate V_M is achieved when the enzyme is saturated with substrate

$$V_M = k_2[E]_T$$

Substituting, yield the Michaelis Menten equation (see Figure 8-1).

$$V = V_M \frac{[S]}{[S] + K_M}$$

Lineweaver and Burk rearranged the equation to give (see Figure 8-2)

$$\frac{1}{V} = \frac{1}{V_M} + \frac{K_M}{V_M} \frac{1}{[S]}$$

Allowing to readily determine the parameters K_M and V_M. More complex reactions can be treated similarly [3].

Effect of Temperature and pH

All enzymes have a temperature optimum and are usually active only in a narrow range of temperatures beyond which they are rapidly dena-

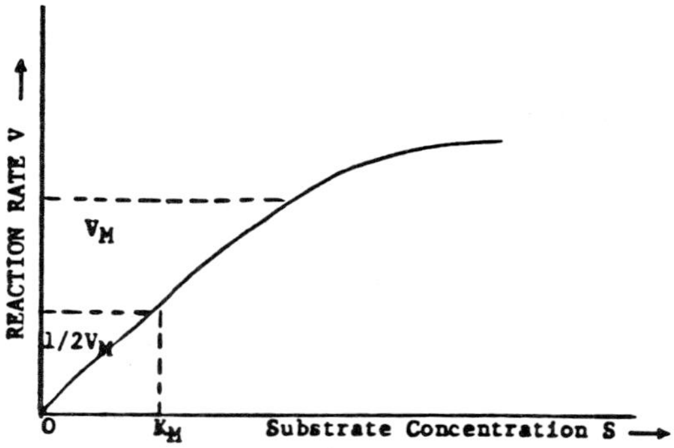

Figure 8-1. Enzyme kinetic.

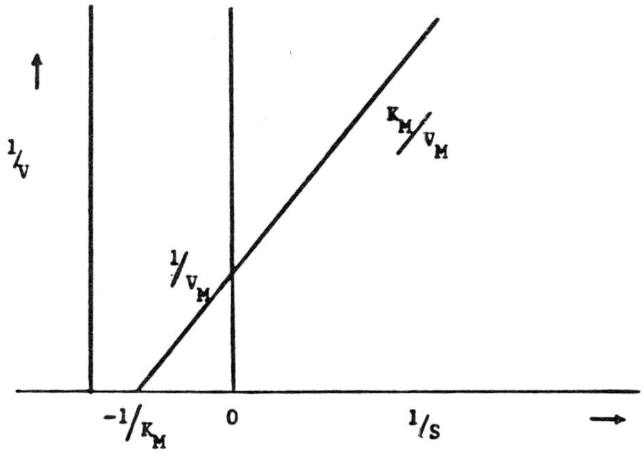

Figure 8-2.

tured and inactivated. Most enzymes are viable between 30 and 80 °C.

Enzyme activity as a function of temperature obeys the Arrhenius equation

$$k = Ae^{'H/RT}$$

Where k is the reaction rate, H the activation energy, T the absolute temperature, R the gas constant and A the Arrhenius constant. Relationships between activity and temperature are known for many enzymes and are typically of the following form (see Figure 8-3).

Most chemical reactions use heat as the source of energy required to initiate the reaction, i.e., to overcome the energy barrier to the reaction. Thermal activation is generally considered to involve the creation of a

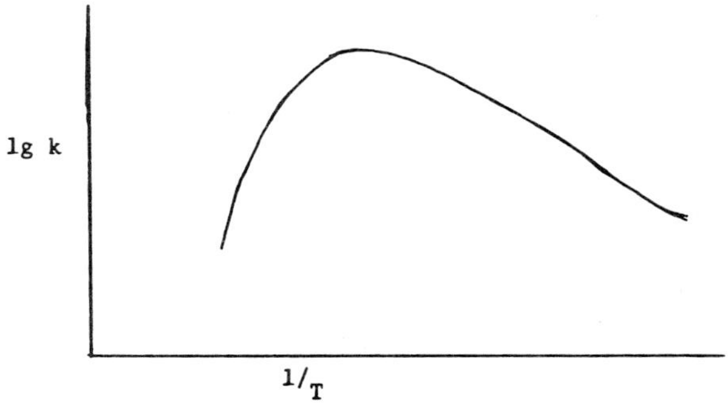

Figure 8-3. Effect of temperature on enzyme activity.

fraction of molecules having energies approximately equal to or greater than the activation energy, E, and can be expressed as:

$$N/N_o = e^{'E/kT}$$

where: N is the number of molecules with energy equal to or greater than E; and N_o is the total number of reactant molecules present; k is the Boltzman constant; and T is the absolute temperature.

The effect of temperature on a chemical reaction is then given by the empirically derived Arrhenius equation.

$$\frac{d \ln k}{dT} = \frac{E}{RT^2}$$

Upon integration we have

$$\ln k = \frac{-E}{R}\frac{1}{T} + \text{constant}$$

Most enzymes follow the Arrhenius plot over a range of physiological temperatures.

Enzyme activity depends strongly on the pH. The pH can affect ionogenic groups of the active site or of neighboring groups; it can change the substrates; modify co-enzymes; it can change the state of non-enzymatic components of the systems; and it can change the shape and structure of the enzyme locally or globally. Many laboratory studies have painstakingly documented these effects for a variety of specific enzyme-substrate combinations. We will not review those here but present a more general picture of the effect of pH on enzymes.

Using the Michaelis formulism, the mathematics of pH effect is straightforward in simple cases of an acid system with two ionisation states [4].

$$AH_2 \underset{K_1}{\rightleftharpoons} AH^- \underset{K_2}{\rightleftharpoons} A^{2-}$$

K_1 and K_2 are the two ionization constants

$$K_1 = \frac{[AH^-]\,[H^+]}{[AH_2]}, \quad K_2 = \frac{[A^{2-}]\,[H^+]}{[AH^-]} = \frac{[A^{2-}]\,[H^+]^2}{[AH_2]\,K_1}$$

$A_s = [AH]_2 + [AH^-] + [A^{2-}]$ is the sum of all the concentrations of the three species.

If AH^- is the only reactive species, the relation between the reaction rate, v and the pH, $[H^+]$ is given by the equations

$$v = k\,[AH^-]$$

$$v = \frac{k\,As}{1 + ([H^+]/K_1) + (K_2/[H^+])}$$

This dependence has a maximum at

$$H^+ = (K_1\,K_2)^{1/2}$$

$$\text{or } pH = 1/2\,(p\,K_1 + p\,K_2)$$

This explains the well-known bell shape curve (Figure 8-4) of the pH versus reaction rate for enzymes.

Many enzymes require the presence of cofactors for full activity. Cofactors include metals and coenzymes. Coenzymes are non-protein organic compounds. Coenzymes are usually involved in reactions dealing with chemical group transfers or with covalent bonds formation. The most commonly encountered coenzymes are ATP, NAD, FAD, coenzyme A, FMN, Biotin and glutathione. Among the metals either bound to enzymes or required for activation, we must mention Zn, Mn, Mg, Cu, Fe, K, Ca, Co, Mn.

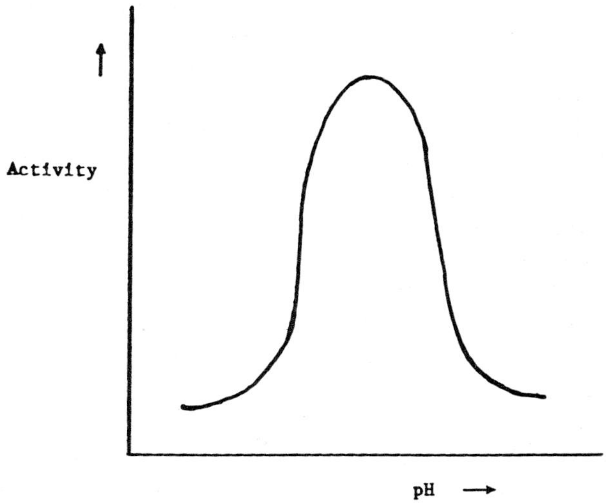

Figure 8-4. Effect of pH on enzyme activity.

ENZYME IMMOBILIZATION

Immobilizing or anchoring enzymes is an important step in extending their applications in industry. There are two categories of immobilized enzymes used in industry: (a) enzymes insolubilized on a water insoluble matrix through adsorption chemical reactions, e.g., insolubilized penicillin amidase can be used to continuously catalyze the hydrolysis of benzyl penicillin; and (b) enzymes contained by an ultrafiltration membrane, e.g., ultrafiltration membrane contains amylase and is used for continuous modification of starch to sugar, that is permeable to reactants and products but not to the enzyme. These two methods have greatly reduced the problems of purification and enzyme removal from end products, and are changing the economic outlook for enzyme-catalyzed reactions.

Progress in the industrial production of pure enzymes has been slow, largely because of the relative instability of most enzymes, the complexity of the isolation procedures, and the cost of the necessary equipment. Once enzymes can be obtained in greater quantities, the scope of their utilization in industry should be greatly increased. One promising approach to reducing cost of enzymes as industrial catalysts is to attach them to an insoluble carrier.

When enzymes are immobilized (e.g., anchored to a glass column or are separated by ultrafiltration) they can catalyze a continuous flow reaction which means that enzymes need no longer be confined to batch reactions. Batch processes are also enhanced by immobilized enzymes anchored to beads which can be easily separated from the final product which leads to two advantages: (1) the enzyme can be reused, reducing cost of catalysis by a factor of perhaps 100; and (2) the finished product is enzyme free.

Four methods for preparing water insoluble enzymes are summarized in Table 8-2. Tanabe Seiyaku Co. of Japan is using immobilized enzymes commercially to catalyze a hydrolysis step in making D- or L-amino acids. Immobilized enzymes are also being studied for a wide variety of purposes such as streamlining the manufacture of synthetic penicillin and the conversion of starches to simple sugars. Other possibilities include chillproofing of beer, inversion of sucrose, desugaring of egg whites and generation of foam in foam rubber. Immobilization allows enzymes to be used in continuous processes and in batch processes. The enzyme-bearing glass beads or cellulosic particles, etc., can be separated quite easily for reuse. A continuous system of converting starch to glucose employs an immobilized enzyme. A method for converting starch to glucose in a continuous system using the enzyme

TABLE 8-2. Enzymes Immobilization.

Immobilization Method	Advantages	Disadvantages
Adsorption		
Physical adsorption on to inert carriers, e.g., glass beads, charcoal, cellulose, or synthetic ion exchange resins.	Inexpensive	Desorption may occur.
Entrapment		
Inclusion inside the lattice of a gel, the pores of which do not allow the entrapped enzymes to escape.	Enzymes unmodified	Enzyme may leak out. Diffusion phenomena affect substrate transport.
Covalent Attachment		
Covalent binding of the enzyme to a porous, wettable, yet water insoluble carrier.	Enzyme unaffected by local environmental parameters.	Enzyme active site may be modified. Expensive.
Covalent Cross Linking		
Covalent cross linking by a bifunctional reagent.	Strong binding	Regeneration not possible.

glucoamylase immobilized on an insoluble carrier has been demonstrated. Most commercial methods for making glucose involve the use of glucoamylase, obtained from *Aspergillus* mold, to convert starch. This is one of the few large-scale industrial uses of enzymes; however, it involves high costs because of the loss of the enzyme itself in the process and loss of the enzyme's activity and stability.

An important subset of the use of immobilized enzymes is that of medical applications of microcapsules. Microcapsules are about the same size as cells, each consisting of a spherical ultra thin membrane (200 Å thick) containing active materials such as enzymes or absorbents. The enveloping membranes, of collodion or nylon, are adjusted to have a functional pore radius of 16 Å, so that although they are impermeable to macromolecules or particulate matter, they are selectively permeable to smaller molecules. Macromolecules or particulate materials inside these artificial cells, while retained in an intracellular environment, can thus act efficiently on any molecules small enough to permeate the membrane, without coming into direct contact with larger molecules from the extracellular environment. Microcapsules may have their most important application in replacement therapy for enzyme defects. The production of these enzymes and their entrapment in microcapsules is an application which may have considerable potential for production of very pure enzymes in small quantities.

THE MARKET

Industrial enzymes are an annual market of some 200 million dollars (Table 8-3). It is not a large market [5]. It is dominated by five products representing 80% of the weight of enzymes sold. Proteases (300 ton/yr), amyloglycosinases (300 ton/yr), amylases (300 ton/yr), glucoisomerase (50 ton/yr) and rennin (10 ton/yr).

We know many more interesting enzymes than are used commercially. In order for an enzyme to be used in industry it must perform a useful function or solve a real problem. The enzyme must be produced and isolated easily from plentiful raw material at a low cost. The enzyme must be acceptable to the industry, the users of the product and the government regulators. The introduction of the enzyme must be made at the right time in the economic cycle of growth, expansion, modernization and retooling. Under these conditions it is surprising how many enzymes have moved from the laboratory to the factory.

Some promising enzymes not yet used in industry or consumer products include rhamnosidase for the debittering of citrus juice, and sulfuydryloxidase enzymes in food processing.

The pulse of the market for industrial enzymes is regularly taken by Wolnak [5] and Hepner [6]. The market is expected to double between 1979 and 1985. Industrial bulk enzymes (like in detergents) represent between less than 1 and 3% of the total product weight. The market in Europe is stronger representing some 50% of the world market. But growth in the U.S. in the next five years will register a higher growth rate. Increased interest in high fructose corn syrup and alcohol production will play a major role in this growth [6].

Enzymes have been harnessed, to the benefit of men, for centuries. Originally enzymes were derived from plant and animal tissues. Today the trend is toward production by fermentation.

ENZYME USES

Enzymes are used in many industrial sectors, consumer products, and scientific pursuits.

TABLE 8-3. Enzymes Market.

				Market—$ Million		
Enzyme	Source	Primary Action	Major Usage	1975	1980	1985 Est.
Amylases						
α,β-amylase	B. subtilis A. oryzae	Hydrolysis of starch	Baking Pulp & paper Textile	8.0	14.8	18.9
amyloglucosidases	A. niger A. oryzae Rhizopus		Beer Alcohol production	6.0	14.3	19.1
Proteases						
Papain	Papaya	Hydrolysis of proteins	Meat tenderizing	10.1	14.9	21.9
Bromelain	Pineapple		Meat tenderizing	1.0	1.3	1.6
Rennin	Animal Mucor miehei (fungus)		Cheese manufacture	14.9	19.9	26.7
Pepsin	Animal		Protein hydrolysates	3.5	4.5	5.8
Trypsin/ Chymotrypsin	Animal		Pharmacy	N/A	8.	8.
Bacterial proteases	B. Subtilis		Detergent Baking	4.7	6.2	8.2
Fungal proteases	B. oryzae		Bread	0.9	1.1	1.4

(continued)

TABLE 8-3. (continued).

				Market—$ Million		
Enzyme	Source	Primary Action	Major Usage	1975	1980	1985 Est.
Lipases	Mucor javanicus A. niger C. lipolytica	Hydrolysis of fats in fatty acids and glycerols	Cheese manufacturing	0.5	0.8	1.3
Celullase	Trichoderma veridae	Hydrolysis of cellulose	Alcohol production	0.3	0.4	0.5
Glucoisomerase	Streptomyces Bacillus	Isomerization of glucose to fructose	HFCS production	15.	50.	65.
Glucooxydase	A. niger	Oxidation of glucose to gluconic acid	Oxygen & glucose removal in food	0.7	1.1	1.3
Pectinase	A. niger	Hydrolysis of pectin	Wine, juice treatment	2.0	2.7	3.6
Invertase	S. Cerevisiae	Hydrolysis of saccharose to glucose and maltose	Invert sugar in candy industry	0.3	0.3	0.3
Lactase	S. Lactis S. fragilis A. niger	Hydrolysis of lactose to glucose and galactose	Lactose hydrolysis in milk	N/A	N/A	4.
Pancreatin	Pancreas		Digestive aids	4.6	5.9	7.5

Source: 5.

Detergents

A major market for enzymes is as detergent. Serine alkaline proteolitic enzymes are the major enzymes in use. They are alkaline stable and active proteases. After an explosion of interest in 1968–1969, enzyme detergents captured a 45% share of the market, the interest was not long sustained and some disappointment led to reduced sales [7,8].

A renewal is taking place associated with the trends toward lower temperature washing, reduced phosphate formulating, and progress in encapsulating technology.

Monitoring and Diagnostic

Enzymes are used extensively in medical diagnosis and are rapidly being incorporated in industrial monitoring programs. In medical diagnosis, enzymes are used as reagents to measure specific metabolites and are measured, as such in body fluids and tissues, as an indicator of pathologic conditions or metabolic disturbances.

The most commonly measured enzymes in clinical chemistry are amylase, alkaline phosphatase, aspartate, aminotransferase, alinine aminotransferase, lactate dehydrogenase, lipase, acid phosphatase, cholinesterase and creatine kinase [9].

Enzyme Reagents and Indicators

A comprehensive survey [10] on enzymes as reagent in clinical medicine is reproduced as Table 8-4. Enzymes used as indicators of physiological or pathological status are listed in Table 8-5.

Enzymes are being used in dual systems such as enzyme electrodes, enzyme thermistors, enzyme mass spectrometer, enzyme electrochemical cells, enzyme polarimeter, enzyme spectrometer and enzyme luminometer [11].

Five important developments are likely to provide a greater role for enzymes in medical and industrial detection and monitoring: DNA probes, enzyme electrodes, enzyme thermistors, enzyme immunoassays and monoclonal antibodies.

DNA Probes

Upward of 2,000 ailments are caused by a single gene defect. Although most such diseases are quite rare (an exception is hypercholesterolemia which hits one in 500 in the U.S.), they add up to about 10% of the total number of diseases affecting us. In a number of cases the se-

TABLE 8-4. Enzymic Methods.

Substances Measured	Enzymes	Detected Substances	Detection Method	Remarks
Acetaldehyde	Alcohol dehydrogenase (EC 1.1.1.1) or	NADH	Photometry	
	Aldehyde dehydrogenase (EC 1.2.1.3)	NADH	Photometry	
Acetoacetate	3-Hydroxybutyrate dehydrogenase (EC 1.1.1.30)	NADH	Photometry	
Acetylcholine	Actylcholinesterase (EC 3.1.1.7)	H^+	Enzyme electrode	
AMP	Adenylate kinase (EC 2.7.4.3)	ADP/ATP	Photometry	DK
		ATP/Photon	Bioluminescence	
cAMP	Phosphodiesterase (EC 3.1.4.1)	ATP/Photon	Bioluminescence	
ADP	Pyruvate kinase (EC 2.7.1.40) and lactate dehydrogenase	ATP/Photon	Bioluminescence	
	(EC 1.1.1.27)	ANDH	Photometry	DK
ATP	Luciferase (EC 1.13.12.7) or	Photon	Bioluminescence	DK,A
	3-phosphoglycerate kinase (EC 2.7.2.3) and glyderaldehyde-3-phosphate dehydrogenase (EC 1.2.1.12)	NADH	Photometry	DK
Alcohol	Alcohol dehydrogenase (EC 1.1.1.1) and	NADH	Photometry	DK,A
	Luciferase (EC 1.14.14.3) or diaphorase (EC 1.6.4.3)	NADH/Photon	Bioluminescence	
	Alcohol oxidase (EC 1.1.3.13)	NADH	Amperometry	
		O_2	Enzyme electrode	
Amino acids	L-Amino acid oxidase (EC 1.4.3.2)	H_2O_2	Photometry	
		H_2O_2	Amperometry	
		$O_2/NH_3/NH_4^+$	Enzyme electrode	
Ammonia	Glutamate dehydrogenase (EC 1.4.1.3) and luciferase	NADH	Photometry	DK,A
	(EC 1.14.14.3)	NADH/Photon	Bioluminescence	
Antithrombin III	Thrombin (EC 3.4.21.5)	p-Nitroaniline	Photometry	DK,A

(continued)

TABLE 8-4. (continued).

Substances Measured	Enzymes	Detected Substances	Detection Method	Remarks
α_1-Antitrypsin	Trypsin (EC 3.4.21.4)	p-Nitroaniline	Photometry	DK,A
Arginine	Arginase (EC 3.5.3.1) and urease (EC 3.5.1.5)	NH_3/NH_4^+	Potentiometry	
Bicarbonate	Phosphoenolpyruvate carboxylase (EC 4.1.1.31) and malate dehydrogenase (EC 1.1.1.27)	NADH	Photometry	DK,A
Bile acids	3α-Hydroxysteroid dehydrogenase (EC 1.1.1.50) and diaphorase (EC 1.6.4.3) or 3β17β hydroxysteroid dehydrogenase (EC 1.1.1.51) and testosterone isomerase (EC 5.3.3.1)	NADH Resorufin Testosterone	Photometry Fluorometry Photometry	DK
Cholesterol	Cholesterolesterase (EC 3.1.1.13) and cholesterol oxidase (EC 1.1.3.6) and peroxidase (EC 1.11.1.7) or	I_3^- H_2O_2 H_2O_2/Photon O_2/H_2O_2	Photometry Photometry Chemiluminescence Amperometry	DK,A DK,A A
	NADH peroxidase (EC 1.11.1.1)	NADH	Photometry	
Citrate	Citrate lyase (ESC 4.1.3.6) and malate dehydrogenase (EC 1.1.1.37)	NADH	Photometry	
Creatine	Creatine kinase (EC 2.7.3.2) and pyruvate kinase (EC 2.7.1.40) and/or lactate dehydrogenase (EC 1.1.1.27) luciferase (EC 1.13.12.7) or creatine amidohydrolase (EC 3.5.2.-) and sarcosine oxidase (EC 1.4.3.-)	NADH ATP/Photon H_2O_2	Photometry Bioluminescence Enzyme electrode	 A
Creatinine	Creatinine deaminase (EC 3.5.4.21) or creatinine amidohydrolase (EC 3.5.2.10) and creatine kinase (EC 2.7.3.2) and pyruvate kinase (EC 2.7.1.40) and lactate dehydrogenase (EC 1.1.1.27) or creatinine amidohydrolase (EC 3.5.2.10) and	NH_3/NH_4^+ NADH	Enzyme electrode Photometry	A DK,A

(continued)

TABLE 8-4. (continued).

Substances Measured	Enzymes	Detected Substances	Detection Method	Remarks
	creatine amidohydrolase (EC 3.5.2.-) and sarcosine oxidase (EC 1.4.3.-)	H_2O_2	Enzyme electrode	A
Creatine phosphate	Creatine kinase (EC 2.7.3.2) and luciferase (EC 1.13.12.7)	ATP/Photon	Bioluminescence	
Dihydroxyacetone	Glycerol kinase (EC 2.7.1.30) and α-glycerophosphate dehydrogenase (EC 1.1.1.8)	NADH	Photometry	
Diphosphoglycerate	Phosphoglycerate mutase (EC 2.7.5.3) and 2,3-diphosphoglycerate phosphatase (EC 3.1.3.13) or 3-phosphoglycerate kinase (EC 2.7.2.3) and/or glyceraldehyde-3-phosphate dehydrogenase (EC 1.2.1.12) luciferase (EC 1.13.12.7)	Phosphate NADH ATP/Photon	Photometry Photometry Bioluminescence	DK DK,A
Estradop; and Estriol	Estradiol-17-β-dehydrogenase and glucose-6-phosphate dehydrogenase (EC 1.1.1.49)	NADPH	Photometry	DK
Fatty Acids	Acyl-CoA synthetase (EC 6.2.1.3) or acyl-CoA oxidase and peroxidase (EC 1.11.1.7) or lipoxidase (EC 1.13.11.12)	AMP H_2O_2 O_2	Photometry Photometry Potentiometry	
Formic acid	Formyl-THF synthetase (EC 6.3.4.3)	5,10-Methenyl-THF	Photometry	DK
Formimino-L-glutamic acid	Formimino-L-glutamic transferase (EC 2.1.2.5) formimino-THF cyclodeiminase (EC 4.3.1.4)	5,10-Methenyl-THF	Photometry	DK
Fructose	Hexokinase (EC 2.7.1.1) and phosphohexose isomerase (EC 5.3.1.9) and glucose-6-phosphate dehydrogenase (ESC 1.1.1.49) or sorbitol dehydrogenase (EC 1.1.1.14)	NADPH NADH	Photometry Photometry	 A
Galactose	Galactose dehydrogenase (EC 1.1.1.48) or galactose oxidase (EC 1.1.3.9)	NADH H_2O_2 H_2O_2	Photometry Photometry Amperometry	DK,A

(continued)

TABLE 8-4. (continued).

Substances Measured	Enzymes	Detected Substances	Detection Method	Remarks
Glucose	Glucose oxidase (EC 1.1.3.4) and peroxidase (EC 1.11.1.7)	H_2O_2	Photometry	DK,A
		H_2O_2/Photon	Chemiluminescence	
		H_2O_2/I^-	Potentiometry	A
		O_2/H_2O_2	Amperometry	DK,A
		O_2/H_2O_2	Enzyme electrode	A
	hexokinase (EC 2.7.1.1) and glucose-6-phosphat dehydrogenase (EC 1.1.1.49) or luciferase (EC 1.13.12.7) glucose	NADPH	Photometry	DK,A
		ATP/Photon	Bioluminescence	
	dehydrogenase (EC 1.1.1.47)	NADH	Photometry	DK,A
Glucose-6-phosphate	Glucose-6-phosphate dehydrogenase (EC 1.1.1.49) and luciferase (EC 1.14.14.3)	NADPH/Photon	Bioluminescence	
Glutamine	Glutaminase (EC 3.5.1.2)	NH_3/NH_4^+	Potentiometry	
		NH_3/NH_4^+	Enzyme electrode	
Glutathione	Glyoxalase (EC 4.4.1.5) and glutathione reductase (EC 1.6.4.2)	NADPH	Photometry	
Glycerol	Glycerol kinase (EC 2.7.1.30) and pyruvate kinase (EC 1.7.1.40) and lactate dehydrogenase (EC 1.1.1.27)	NADH	Photometry	
Glycerophosphatides	Phospholipase C (EC 3.1.4.3)	Glycerol	Photometry	
Heparin	Thrombin (EC 3.4.21.5)	p-Nitroaniline	Photometry	DK
3-Hydroxybutyrate	3-Hydroxybutyrate dehydrogenase (EC 1.1.1.30)	NADH	Photometry	
2-Ketoglutarate	Glutamate dehydrogenase (EC 1.4.1.3)	NADH	Photometry	
Lactate	Lactate dehydrogenase (EC 1.1.1.27)	NADH	Photometry	DK,A
		Fe^{2+}/Fe^{3+}	Potentiometry	
		Fe^{2+}/Fe^{3+}	Enzyme electrode	

(continued)

TABLE 8-4. (continued).

Substances Measured	Enzymes	Detected Substances	Detection Method	Remarks
Lecithin	Phospholipase D (EC 3.1.4.4) and choline oxidase (EC 1.1.3.17) and peroxidase (EC 1.11.1.7)	H_2O_2 H_2O_2	Amperometry Photometry	 DK,A
Lipoprotein-X	Cholesterol oxidase (EC 1.1.3.6) and peroxidase (EC 1.11.1.7)	H_2O_2	Photometry	
α_2-Macroglobulin	Trypsin (EC 3.4.21.4)	p-Nitroaniline	Photometry	DK
Methionine	Methioninelyase (EC 4.4.1.11)	NH_3/NH_4^+	Potentiometry	
Ornithine	Ornithine carbamoyl transferase (EC 2.1.3.3)	Citrulline	Photometry	
Orthophosphate	Phosphorylase (EC 2.4.1.1) and phosphoglucose mutase (EC 2.7.5.1) and glucose-6-phosphate dehydrogenase (EC 1.1.1.49)	NADPH	Photometry	DK,A
Oxalate	Oxalate decarboxylase (EC 4.1.1.2) or oxalate oxidase (EC 1.2.3.4) and catalase (EC 1.11.1.6) and aldehyde dehydrogenase (EC 1.2.1.3)	CO_2 H_2O_2/Photon NADH	Conductometry Chemiluminescence Photometry	
Pyrophosphate	Uridine-5′-diphosphoglucose pyrophosphorylase (EC 2.7.7.9) and Phosphoglucose mutase (EC 2.7.5.1) and glucose-6-phosphate dehydrogenase (EC 1.1.1.49)	NADPH	Photometry	A
Pyruvate	Lactate dehydrogenase (EC 1.1.1.27) and luciferase (EC 1.14.14.3)	NADH NADH/Photon	Photometry Bioluminescence	DK,A
Sphingomyelin	Spingomyelinase (EC 3.1.4.12) and alkaline phosphatase (EC 3.1.3.1) and choline oxidase (EC 1.1.3.17) and peroxidase (EC 1.11.1.7)	H_2O_2	Photometry	DK
Steroids	β-Glucuronidase (EC 3.2.1.31) and arylsulphatase (EC 3.1.6.1) and 3αhydroxysteroid dehydrogenase (EC 1.1.1.50) and/or diaphorase (EC 1.6.4.3) peroxidase (EC 1.11.1.7)	Hydroxysteroids NADH Formazan H_2O_2	Photometry Photometry Photometry Photometry	

(continued)

TABLE 8-4. (continued).

Substances Measured	Enzymes	Detected Substances	Detection Method	Remarks
Triglycerides	Lipoprotein lipase (EC 3.1.1.23) and glycerol dehydrogenase	NADH	Photometry	
	(EC 1.1.1.6) and diaphorase (EC 1.6.4.3) or lipase (EC 3.1.1.3)	Formazan	Photometry	DK,A
	and chymotrypsin (EC 3.4.21.1) and glycerol kinase			
	(EC 2.7.1.30) and/or pyruvate kinase (EC 2.7.1.40) and lactate			
	dehydrogenase (EC 1.1.1.27) or -blycerophosphate dehydrogen-	NADH	Photometry	DK,A
	ase (EC 1.1.1.8) and diaphorase (EC 1.6.4.3) or glycerol kinase	Formazan	Photometry	DK,A
	(EC 2.7.1.30) and luciferase (EC 1.13.12.7)	ATP/Photon	Bioluminescence	
Tyrosine	Tyrosine decarboxylase (EC 4.1.1.25)	CO_2	Potentiometry	
Urea	Urease (EC 3.5.1.5)	NH_3/NH_4^+	Photometry	DK,A
		NH_3/NH_4^+	Enzyme electrode	A
		NH_3/NH_4^+	Potentiometry	
		H^+	Potentiometry	A
		NH_4^+ HCO_3^-	Conductometry	DK,A
Urea	Urease (EC 3.5.1.5) and glutamate dehydrogenase (EC 1.4.1.3)	NADH	Photometry	DK,A
Uric acid	Uricase (EC 1.7.3.3)	Uric acid	Photometry	DK,A
		O_2	Amperometry	A
		CO_2	Potentiometry	
	Uricase (EC 1.7.3.3) and	H_2O_2/Photon	Chemiluminescence	
	peroxidase (EC 1.11.1.7) or catalase (EC 1.11.1.6) and	H_2O_2	Photometry	DK,A
	aldehyde dehydrogenase (EC 1.2.1.5)	NADPH	Photometry	DK,A
Xanthine and Hypoxanthine	Xanthine oxidase (EC 1.2.3.2)	Purine nucleotides	Photometry	
Xylose	Xylose dehydrogenase (EC 1.1.1.-)	NADPH	Photometry	

DK = Diagnostic kits, commercially available
A = Automated assays
Source: 10.

TABLE 8-5. Enzymes Measured for Clinical Utility.

Acetylcholinesterase	Hexosephosphate Isomerase
Acid Phosphate	Hexose-1-Phosphate Uridyl Transferase
Alanine Aminotransaminase	—Hydroxybiturate Dehydrogenase
Aldolase	Isocitrate Dehydrogenase
Alkaline Phosphate	Lactate Dehydrogenase
Aminolevulinate-Dehydratase	Leucine Aminopeptidase
Amylase	Lipase
Arginase	Lipoprotein Lipase
Arginosuccinate Lyase	Lysozyme
Aspartate Aminotransaminase	Malate Dehydrogenase
Ceruloplasmin	NADH Methemoglobin Reductase
Cholinesterase	5′-Nucleotidase
Chymotrypsin	Ornithine Carbamoyl Transferase
Creatin Kinase	Pepsin & Pepsinogen
Cytochrome C Oxidase	Phosphoglucomutase
Galactose-1-Phosphate Uridyl Transferase	Plasmin
Gamma Glutamyl Transferase	Pyruvate Kinase
Glucose-6-Phosphate Dehydrogenase	Sorbitol Dehydrogenase
Glucuronidase	Triosephosphate Isomerase
Glutathionine Reductase	Trypsin
Guanase	Uropepsin

Source: 9.

quence of the gene involved is well known and the mutated base pairs well identified (e.g., β^{39}-globin in thalassaemia). It is then possible to design a DNA probe suitably labelled to hunt for such markers. This is already possible for such diseases as sickle cell anaemia, thalassaemia, and Lesch-Nyhan syndrome. A form of diarrhea-causing bacteria has been detected by a DNA probe in human samples. Possible in the very near future would be the detection of haemophilia (factor VII), coagulation factor IX, phenylketonuria, Duchenne muscular dystrophy, and Huntington's chorea. Most rewarding (technically and fiscally) would be a DNA probe to detect hypertension (the Hyp-1 gene producing an enzyme regulating the production of 18-hydroxy-deoxy-corticosteroid has been identified at the Medical College of Ohio), or alcoholism (the gene for the enzyme responsible for the accumulation of acetaldehyde). The market in this area is wide open.

Enzyme Electrodes

More than 60 enzyme electrode probes have been described in the literature. Many are totally specific for a single chemical. The substrate measured include sugars (glucose, sucrose, maltose, lactose), acids (acetic, uric, formic, oxalic, lactic), amino acids (arginine, aspargine, cysteine, glutamine, glutamic acid, histidine, lysine, phenylalanine, tyrosine), urea, cholesterol, creatinine, penicilin and acetylcholine [12–16].

Microbial electrode sensors have been developed. The whole cell is immobilized and a product of the reaction is measured. A few examples include [17,18]:

Substrate	Immobilized Microorganisms	Sensor
BOD	*Trichosporon cutaneum*	O_2 probe
Ammonia	Nitrifying bacteria	O_2 probe
Acetic acid	*Trichosporon brassiere*	O_2 probe
Ethyl alcohol	*Trichosporon brassiere*	O_2 probe
Glutamic acid	*Escherichia coli*	CO_2 probe

Ion-selective electrodes are increasingly used to measure pH, gas, ions, enzymes and organic substrates. They are small, rugged and accurate.

The response of an ion selective electrode to an ion A of activity a and charge Z in the presence of an interfering ion B is given by

$$E = \text{const.} + \frac{2.303\ RT}{Z_A\ F}\ \lg\ [a_A + K_{pot}^{A,B}\ (a_B)Z_A/Z_B]$$

Where E is the measured potential, R is the gas constant (8.314 JK^{-1}), T is the temperature (°K), F is the faraday constant (96,487 °C Mol^{-1}) and $K_{pot}^{A,B}$ is the selectivity constant.

There are several types of ions selective electrodes.

Glass Electrodes

The composition of the glass determines the selectivity. Glass electrodes are available to measure pH and cations concentrations.

Membrane Electrodes

A selective membrane permeable to ions or gases which permeates the membrane and alters the pH of an internal filling solution can be made selective to very specific species in the presence of interferences.

Enzyme Electrodes

The electrode can be covered with a membrane supporting an immobilized enzyme. The electrode detects a product of the enzyme/substrate reaction which can be correlated with substrate concentration. Glucose oxidase so immobilized will measure glucose concentration through a pH change, the diappearance of O_2 on the production of oxygen.

Guilbault [12–16] and others [17–24] have summarized the available enzyme electrode probes. The current list is given in Table 8-6.

TABLE 8-6. Enzyme Electrodes.

Substrate	Enzyme	Sensor
Acetic & formic acids	Alcohol oxidase	
Acetylcholine	Acetylcholinesterase	Choline
Acetyl-β-methylcholine	Acetylcholinesterase	Acetylcholine
Adenosine monophosphate	5-Adenylate deaminase	NH_4^+
Alcohols†	Alcohol dehydrogenase	Pt
	Alcohol dehydrogenase/diaphorase	Pt
	Alcohol oxidase	Pt
D-Amino acids*	D-Amino acid oxidase	NH_4^+
L-Amino acids** (general)	L-Amino acid oxidase	Gas (NH_3) NH_4^+ Pt
	Decarboxylases	CO_2
L-Arginine	Arginase	NH_4^+
L-Asparagine	Asparaginase	NH_4^+
L-Cysteine	*Proteus morganii*	H_2S
L-Glutamine	Glutaminase	NH_4^+
L-Glutamic acid	Glutamate dehydrogenase	NH_4^+
	Glutamate decarboxylase	CO_2
L-Histidine	Histidinase	NH_4^+
L-Lysine	Lysine decarboxylase	CO_2
L-Methionine	Methionine ammonia lyase	NH_3
L-Phenylalanine	Phenylalanine ammonia lyase	NH_3
L-Tyrosine	Tyrosine decarboxylase	CO_2
	Tyrosinase	Gas(O_2)
Amygdalin	β-Glucosidase	CN^-
Butyryl thiocholine	Cholinesterase	Pt(SCh)

(continued)

TABLE 8-6. (continued).

Substrate	Enzyme	Sensor
Cholesterol	Cholesterol esterase/cholesterol oxidase	Pt(H_2O)
		Pt(O_2)
	Cholesterol oxidase	Pt(O_2)
Creatinine	Creatininase	NH_4^+
	Creatininase (purified)	NH_3
Glucose	Glucose oxidase	pH
		pt(H_2O_2)
		Pt(Quinone)
		Pt(DCIP)
		Pt(O_2)
		I^-
		Gas(O_2)
	Glucose oxidase peroxidase	Pt
Lactic acid	Lactate dehydrogenase	Pt^-
		$Fe(CN)_6^{4-}$
		C(NADH)
	Cyt-b_2	Pt
Lactose	β-Galactosidase/glucose oxidase	Gas (O_2)
Lectin	Glucose oxidase	Gas (O_2)
Maltose	Maltase/glucose/oxidase	Gas (O_2)
NADH	Alcohol dehydrogenase	Pt
	Mitochandria	Gas (O_2)
Nitrate	Nitrate reductase/nitrite reductase	NH_4^+
Nitrite	Nitrite reductase	Gas (NH_3)
Oxalic Acid	Oxalate decarboxylase	Gas (CO_2)
Penicillin	Penicillinase	pH
Peroxide	Catalase	Pt(O_2)
Phenol	*Trichosporon cutaneum*	Gas (O_2)
Phosphate	Phosphatase/glucose oxidase	Pt(O_2)
Saccharose	Invertase/mutarotase/glucose oxidase	Pt(O_2)
Succinic acid	Succinate dehydrogenase	Pt (O_2)
Sucrose	Sucrase/glucose oxidase	Pt(H_2O_2)
Sulfate	Aryl sulfatase	Pt
Thiosulfate	Rhodanase	CN^-
Urea	Urease	NH_4^+

(continued)

TABLE 8-6. (continued).

Substrate	Enzyme	Sensor
		pH
		Gas(NH_3)
		Gas(CO_2)
Uric acid	Uricase	Pt(O_2)

*Responds to D-phenylalanine, D-alanine, D-valine, D-methionine, D-leucine, D-norleucine, and D-isoleucine.
**Responds to L-leucine, L-tyrosine, L-phenylalanine, L-tryptophan, and L-methionine.
†Responds to methanol, ethanol, allyl alcohol.
Source: 14.

Enzyme Thermistors

A related concept is the enzyme thermistor. When a substrate is fed to an enzyme, a product is formed with liberation of heat. A thermistor is used to monitor this heat of reaction which can be correlated with the concentration of substrate [25–27].

Enzyme thermistors have been developed to measure ethanol, cephalosporin, galactose, lactose, sucrose and penicilin G [28,29].

Enzyme Immunoassays

Enzyme immunoassays is rapidly replacing radioactive immunoassay. A series of careful studies has placed this technique on a firm foundation [30–37].

The use of enzyme label in immunoassay was introduced in 1971. Many variants of the technique have been tested [38–44]. In its simplest form, the method works as follows:

1. A labeled antigen competes with unlabeled antigen for binding to an antibody.
2. The antibody-bound antigen is separated from the free antigen by the use of a solid phase antibody or a second antibody.
3. The enzyme activity of the bound or free faction is determined.

The major advantages of the techniques are 1) stability of the label, 2) sensitivity, 3) selectivity, 4) speed, 5) reduced cost, 6) ease of detection by photometry or with an electrode, 7) and the avoidance of a separation step in some cases. This technique has been used to measure antigens/haptens and antibodies. The technique can also be used in drug screening of urine (Table 8-7) [45–48].

TABLE 8-7. Applications of Enzyme Immunoassay in Diagnosis.

Substance to be Measured	Label	Separation Technique	Substance Detected
Albumin, serum	Glucoamylase	Solid phase (agarose)	
Anti-adenosine antibody	β-galactosidase		o-nitrophenol
Anti-DNA antibody	peroxidase	solid phase (coated tube)	H_2O_2
Antibodies to *Salmonella* O antigens	alkaline phosphatase	solid phase (coated tube)	
Estrogens (total)	peroxidase	double-antibody solid phase	H_2O_2
α-Fetoprotein (rat)	alkaline phosphatase	solid phase (coated tube)	p-nitrophenol
α-Fetoprotein (human)	peroxidase	double antibody	H_2O_2
	glucose oxidase	solid phase	
	alakline phosphatase	double antibody	
Haptoglobin	alkaline phosphatase	solid phase (coated tube)	p-nitrophenol
Chorionic gonadotropin (human)	peroxidase	solid phase (cellulose)	H_2O_2
	peroxidase	double-antibody solid phase	H_2O_2
α_2H-globulin	peroxidase	double antibody	H_2O_2
Placental Lactogen (human)	peroxidase	double-antibody	H_2O_2
IgE	alkaline phosphatase	solid phase (coated tube)	p-nitrophenol
IgG	alkaline phosphatase	solid phase (coated tube)	p-nitrophenol
	peroxidase	solid phase (Biogel)	H_2O_2
	glucose oxidase	solid phase (Sepharose)	
Insulin	glucose oxidase	solid phase (Sephadex)	
	peroxidase	solid phase (coated tube)	H_2O_2

(continued)

TABLE 8-7. (continued).

Substance to be Measured	Label	Separation Technique	Substance Detected
Insuline	alkaline phosphatase		phenylphosphate
Estradiol, estriol	peroxidase	double-antibody solid phase	H_2O_2
Pregnancy-associated α-macroglobulin	peroxidase	solid phase (coated tube)	H_2O_2
Thyroxine	malate dehydrogenase		NADH
Tyroxine	Peroxidase		H_2O_2
Carcinoembryonic antigen	Alkaline phosphatase	solid phase (coated tube)	p-nitrophenol
Cortisol	β-galactosidase	Double antibody	o-nitrophenol
	alkaline phosphatase		p-nitrophenol
	glucose-6 P-dehydrogenase		NADPH
2,4 dinitrophenol	Alkaline phosphatase	Solid phase (polystyrene tube)	p-nitrophenol
Tyrotropin	Alkaline phosphatase		Phenylphosphate
Tyrotropin	Peroxidase	Double antibody	H_2O_2
Ferritin	Alkaline phosphatase		pl-nitrophenol
Tridothyronine	Peroxidase		H_2O_2
I_gG/I_gM	β-galactosidase		o-nitrophenol
Insuline	β-galactosidase		4-methyl-umbelliferone
Progesterone	β-galactosidase	Solid phase cross linked protein	o-nitrophenol
Hepatitis B surface antigen	Peroxidase	Solid phase	H_2O_2
Vibrio cholerae	Alkaline phosphatase	Solid phase (polystyrene tube)	
extoxin	Alkaline phosphatase	Solid phase (polystyrene tube)	
Lipopolysaccharide	Alkaline phosphatase	Solid phase (polystyrene tube)	

Monoclonal Antibodies

Of all the biotechnologies, the production and use of monoclonal antibodies (MAB) is the most rapidly expanding field and is reaching the marketplace faster than any other product. The production of monoclonal (pure clones) antibody is rather simple. An antigen (a foreign substance to an organism) of interest is injected into a mouse. A few days later the mouse's spleen is removed. It contains large amounts of white cells producing antibodies to the antigen injected. These cells are mixed with fast dividing cancer cells—myelomas and a chemical, polyethylene glycol (to effect fusion between the two kinds of cells). The result is an hybridoma. The fusion of the antibody producing cells with the fast dividing cancer cells provides a product retaining both capabilities: virtually dividing forever and producing an endless supply of pure antibodies. The hybridoma is usually kept alive in a mouse which is tapped at regular intervals to retrieve material from which the antibodies are separated. *In vitro* culturing has also been applied to the production of monoclonal antibodies. It is possible to produce MAB against any specific antigen. The specificity, speed, and sensitivity of the antigen antibody reaction make it an ideal detector for complex molecules present in small quantity in a mixture of other chemical species. MAB can be engineered for detecting infectious agents, viruses, cell surface markers, hormones, and chemicals in general. MAB for some 200 antigens are commercially available mostly for research purposes. The detection of the antigen/antibody reaction can be effected with a color reaction (dye), a radioactive market, an enzyme electrode, or by simply observing agglutination.

The technique for producing monoclonal antibodies *in vivo* and *in vitro* is well established. More than 200 MAB systems are available commercially for research and some 50 MAB kits are now on the market or are close to FDA approval.

Immunodiagnostics are now 20% of a $250 million U.S. medical diagnostic industry. The *in vitro* diagnosis market is estimated at $1 billion in 1990, 50% of all the diagnostic tests will be based on monoclonal antibodies.

Some of the commercially available, or soon to be available, MAB systems include:

Prostatic Acid Phosphatase (PAP)	Abbott Lab., Hybritech
Pregnancy test (hCG Hormone)	Monoclonal Antibodies, Inc.
T and B lymphocytes, monocytes, leukocyte	Johnson & Johnson

Viral Hepatitis	Centocor, Inc.
Pancreatic, gastric, colorectal cancers	Centocor, Inc.
Carcinoembryonic antigen (CEA)	Abbott Lab.
Plant viruses	American Type Culture Collection
Ferritin	Hybritech
Immunoglobulin E	Hybritech
Prolactin	Hybritech
Growth hormone	Hybritech
Rheumatic fever susceptibility	Rockefeller Univ.
Protein Myosin	Mass. Gen. Hosp.
Hodgkins disease	Albrecht Univ., FRG
Salmonella	Immunocell Corp.
Respiratory diseases diagnosis Pneumonia Cold Legionnaire's disease	Genetic Systems Corp.
Gonorrhea, syphilis, cytomegalovirus, varicella zoster, chlamydia, herpes	Genetic Systems Corp.
Different human blood cells	Genetic Systems Corp.
Anti-A and anti-B	Celltech, Ltd.
Leukemia, breast, lung, colon, prostate cancers	Genetic Systems, Oncogen
Radioactive MAB (cancer treatment)	Johns Hopkins, Hybritech, Inc.
Fatal diarrhea in calves	Molecular Genetics
Radioactive MAB (cancer diagnosis)	Nuclear & Genetic Technology
Human T cell leukemia virus	Biotech Research Lab.
Human proteins (Ig's, etc.)	Serotex

Food Industry

The largest commercial use of enzymes is in the production of high fructose corn syrup (HFCS). High fructose corn syrup is a sweetener, an order of magnitude higher than sucrose. It is produced from corn through microbial enzymes—glucoamylase and glucose isomerase that converts starch to glucose and glucose to fructose.

A large number of enzymes are used in food processing for fermentation, sugar conversion, oxygen removal, fat modifications, protein digestions, oxido reduction reactions, alcohol production, flavor development and enhancement, etc. Single cells proteins are produced around the world for animal feed and meat extender [49].

The Food and Drug Administration requires data on the food use of enzymes in order to regulate those substances. A 1978 survey of manufacturers and users of enzyme products was conducted [50]. Some enzymes (soluble enzymes) are retained in food products after processing. The FDA defines a maximum allowable level. Table 8-8 is derived from the survey mentioned above.

Chemicals

A number of important industrial chemicals are produced by fermentation: Acetone/butanol (from starchy material), ethanol (from sugar/starch/cellulose), acetic acid (from alcohol), citric acid (from sugar/starch), gluconic acid (from glucose), itaconic acid (from molasses), lactic acid (from carbohydrates), and polysaccharides such as xanthan gum, and dextran.

Pharmaceutical

The major commercial route for making cortisone is from stigmasterol which is chemically converted to progesterone. Hydroxylation of progesterone is done by *Rhizophus nigricans* which yield 11-α-hydroxy progesterone.

Cholesterol is easily degraded in steps by bacteria. The homoallylic alcohol in cholesterol is oxydized to the enone and dehydrogenated to the dienone. Side chain cleavage form the carboxylic acid which is cleaved to the androstadiendione which can be hydrogenated to androstenedione.

Microbial transformation of steroids is important to the pharmaceutical industry. These are all enzymatic reactions involved in fermentation vats with the aid of live microorganisms [1].

Substrate	Product	Organism
Cholic acid	Ketocholic acid	*Alcaligenes faecalis*
Progesterone	11 α-hydroxyprogesterone	*Rhizopus nigricans*
Cortisol	Prednisolone	*Arthrobacter simplex*
Progesterone	1-dehydrotestololactone	*Cylindrocarpon radiciola*
β-sitosterol	Andaustadienedione	*Mycobacterium*
Oestranol	Oestrone	*Streptomyces albus*

The classic process among pharmaceuticals is the production of penicilins G and V by mutants of *Penecillium chrysogenum*. A number

TABLE 8-8. Enzymes Used in Food.

	α-amylase	β-amylase	Bromelain	Catalase	Cellulase	Dextranase	Ficin	β-glucanase	Glucoamylase	Glucose isomerase	Glucose oxidase	Hemicellulase	Invertase	Lactase	Lipase	Pancreatin	Papain	Pectinase	Pepsin	Protease	Pullunase	Rennet	Trypsin
01 Baked goods/baking mixes	X		X		X							X			X		X	X		X			
02 Breakfast cereals	X														X				X				
03 Other grain products, pastes	X																						
04 Fats and oils	X		X		X						X	X			X					X			
05 Milk, milk products				X	X							X		X	X			X	X			X	
06 Cheese			X	X											X	X			X	X		X	X
07 Frozen dairy desserts					X							X		X	X			X		X			
08 Fruit and fruit juices					X				X		X	X						X					
09 Fruit ices, water ices									X														
10 Meats, meat products			X				X													X			
11 Poultry, poultry products																	X			X			
12 Eggs, egg products				X							X				X								

(continued)

TABLE 8-8. (continued).

	α-amylase	β-amylase	Bromelain	Catalase	Cellulase	Dextranase	Ficin	β-glucanase	Glucoamylase	Glucose isomerase	Glucose oxidase	Hemicellulase	Invertase	Lactase	Lipase	Pancreatin	Papain	Pectinase	Pepsin	Protease	Pullunase	Rennet	Trypsin
13 Fish, seafood																							
14 Vegetables, vegetable juices	X				X				X			X											
15 Condiments, relishes									X									X					
16 Candy	X		X									X	X		X					X			
17 Sugars, frostings	X	X			X	X			X	X				X	X			X		X	X		
18 Jam, jellies, sweet spreads																		X					
19 Sweet sauces, toppings, syrups	X		X						X									X		X			
20 Gelatines, puddings, custards	X		X		X							X			X		X			X		X	
21 Soup, soup mixes																		X					
22 Snack foods			X												X		X			X			
23 Beverages (non-alcoholic)											X	X			X			X					

(continued)

TABLE 8-8. (continued).

	α-amylase	β-amylase	Bromelain	Catalase	Cellulase	Dextranase	Ficin	β-glucanase	Glucoamylase	Glucose isomerase	Glucose oxidase	Hemicellulase	Invertase	Lactase	Lipase	Pancreatin	Papain	Pectinase	Pepsin	Protease	Pullunase	Rennet	Trypsin
24 Beverages (alcoholic)	X	X			X		X	X	X		X						X	X		X			
25 Nuts, nut products																							
26 Reconstituted veg. proteins			X												X	X	X		X	X		X	
27 Gravies, sauces	X		X				X								X		X			X			
28 Dairy product analogs			X												X				X	X		X	
31 Chewing gum			X																	X			
45 Main dishes																							
46 Food colors																							
47 Dietary supplements																							
48 Seasonings and flavorings			X				X					X			X		X	X		X			
49 Coffee and tea				X							X	X						X					
82 Baby food cereals	X																						

Source: 50

of vitamins are produced by bacterial action, including cyanocobalamin (B_{12}), and riboflavin (B_2).

Medical Applications

The current therapeutic use of enzymes are very few but it is a promising field likely to benefit greatly from a rapidly growing knowledge of enzymology and the availability of many new enzymes with specialized properties [51,52].

Enzymes have been used in a solubilized form or immobilized on some substrate; they have been given orally, injected or implanted in soft tissue as prostheses; and they also have been used in extracorporeal equipment.

To be used *in vivo*, enzymes must meet stringent requirements:

- The enzyme must be larger than 60,000 daltons to avoid immediate glomerular filtration and excretion in urine
- They must be protected through encapsulation or immobilization to avoid denaturation and proeolysis
- They must not activate immunological reactions
- They must have high purity to insure efficacity and avoid side effects
- Targeted application
- Control over duration of physiological activity [53].

A few examples should give an idea of the range of possibilities of enzymes in therapy.

Hyaluronidase—this enzyme hydrolyses hyalurinic acid, polysaccharide. The enzyme is used as a dispersion agent with injected drugs.

L-asparaginase—the enzyme is promising in the treatment of lymphoid leukosis. Certain kinds of leukemia require an exogenous source of asparagine for growth. The enzyme, L-asparaginase acts by hydrolyzing L-asparagine in the plasma, achieving tumor mass reduction.

Digestive aid enzymes—these enzymes (a mixture of enzymes) are largely used in Japan to aid digestion.

Fibrinolytic Agents

The obstruction of a blood vessel by a blood clot can have catastrophic effect. Thrombos have formed in the arterial and venous systems under a variety of conditions. The protein fibrin serves as a matrix to support the clots. The intent of fibrinolysis is to proteolytically degrade fibrin to a soluble product. The human fibrinolytic system is complex and fine tuned. Strektokinase and urokinase are two enzymes involved in the activation of the endogeneous plasminogen system controlling the proteolytical action of plasmin on fibrin. Both enzymes have been used in the treatment of patients with blood clots [54–57].

Anti-inflammatory Agents

Proteolytic enzymes have been used for the reduction of inflammation and edema. Their beneficial action is associated with the solution of fibrin and the clearing of proteinaceous debris. Bromelain, pancreatin, collagenase among others, have been used in this mode.

Collagen

Collagen is the major component of connective tissue. The physical and chemical nature of collagen can be readily modified. It is possible to cast collagen into a variety of shapes or into diverse surfaces. Collagen would be an ideal candidate for prostheses such as cornea, heart valves, blood vessels, etc. [58]. It has undergone tests as dialysis membranes and has proven beneficial in healing dermal wounds.

Other Applications

Enzymes can be used as detoxifying agents against barbiturate overdose or cyanide poisoning, as substitution to correct genetic metabolic error (such as phenylketouria) and in the synthesis of lacking metabolites [59].

Future Prospects

More complex immobilized and encapsulated multienzyme systems are the answer to many therapeutic needs. Additional oncolytic enzymes (similar to *L. asparginase*) are under development or test (Neuraminidase, ribonoclease, carboxypeptidase, L. methioninase, L. tyrosinase, etc.) [1].

ENZYMES IN RESEARCH

A class of enzymes commercially available in pure form but in small quantities are restriction endonucleases. These cutting enzymes are the scissor used in industry and laboratories to snip open DNA segments—the first step in all DNA recombinant techniques.

A number of other purified enzymes are used in small quantity in research and experiments. A number of specialized technical laboratories provide these enzymes to investigators. Ribonuclease, polymerase, etc.

SURVEY OF ENZYMES USED IN INDUSTRY

Enzymes used in industrial operations are rarely the pure product. More often they are a crude preparation with sufficient activity to perform the intended function. Often enzymes are used as a mixture or in combination to achieve multiple purposes simultaneously.

The attached table (Table 8-9) is a compilation of reported uses of enzymes from many sources. The intent was to be as comprehensive as possible in a rapidly changing business sector.

An important group of enzymes is used as reagents in clinical diagnosis and industrial measurement systems and a vast number are used in research. These are not well reported in the table. Table 8-9 is derived from a variety of sources [1,60,61].

PERSPECTIVE

The future for the industrial use of enzymes is bright. Projecting the immediate past into the near future allows to make a few safe predictions.

- Enzymes will be used more extensively in the near future in detergents, food processing, and pharmaceutical products development.
- Enzymatic photography will be tested in the marketplace [62].
- One hundred percent high fructose corn syrup (HFCS) based on glucose 2 oxidase will replace much of today's 42% and 55% HFCS.
- Vast stores of available lignin (from the pulp industry) will be converted to phenols, cresols and other valuable chemicals.
- Enzymes will play a key role in developing biorefineries to process waste material and biomass to alcohol and other chemicals.
- Enzymes will be called upon to serve new needs in medical, diagnostic and industrial monitoring.

None of these forecasts is a strong departure from the present. The economic balance sheet will dictate the direction and size of the market [63].

TABLE 8-9. Enzymes Survey.

Principle Enzyme	Additional Enzymes	Primary Action on Substrate	Utilization
A			
α-amylase		liquefies starch	Production of adhesives, remove starch sizings from textiles, paper coatings, conveyor belt cleaning, baking industry (reduction in dough viscosity), sewage disposal, food ingredients (bread, baby food), adhesives, starch, hydrolyzes converted starch syrups to dextrose syrups, pharmaceuticals, digestive aids, feedstuffs for poultry, beef, cattle, swine, wall-paper removal
	α-amylase, protease, phosphate, maltase	hydrolyzes starch	fruit juices, chocolate syrup, brewing
β-amylase		hydrolyzes starch	maltose production in brewing
	α-amylase	converts starch to maltose	brewery, distillery, bakery
	β-amylase	hydrolyzes starch	baking
		liquefies and saccharifies starch	baby foods, pharmaceuticals, sugar, syrups, brewing
	α-amylase		
	β-amylase, amyloglucosidase	hydrolyzes starch, dextrins	syrup production from acid hydrolyzed starch
	α-amylase, heat-stable pentosanase	liquefies starch and substituted starches pentosans	baking, brewing, cereals, cocoa and candy, distillation and fermentation, laundry, paper, starches and syrups, textiles
	α-amylase, protease	liquefies starch	textiles, adhesives, paper, brewing, grain alcohol, baking
	α-amylase,	liquefies and saccharifies	pharmaceuticals

(continued)

TABLE 8-9. (continued).

Principle Enzyme	Additional Enzymes	Primary Action on Substrate	Utilization
	β-amylase, protease	starch	
	amylase, protease lipase	hydrolyzes proteins, starch and fat, microorganisms	all-purpose digestive enzyme, pharmaceuticals research, leather, textile desizing
	amylase, protease	liquifies starch	adhesives, sizing paper coatings, textile desizing, waste disposal
	amylase, protease, lipase, and anaerobic bacteria	liquefies sewage	cleaning drain lines, grease traps, sewage treatment
	β-amylase, protease amyloglucosidase	converts dextrins to maltose and dextrose	cereal syrups—higher DE
	α-amylase, amylopsin, animal diastase	hydrolyzes starch	digestive aid, pharmaceuticals
	amylase, lipase	hydrolyzes starch, fat	food products
	amylase, protease	hydrolyzes starch and proteins	crackers, wafers
	α-amylase, protease	hydrolyzes starch	desizing of textiles
	α-amylase (heat treatment)	hydrolyzes starch	starch, syrups with low sugar content
amyloglucosidase		hydrolyzes starch to glucose	dextrose manufacture, sweetener, fermentation, pharmaceuticals, convert oligosaccharides to glucose in light beer
	amyloglucosidase, β-amylase	saccharification of starch	production of dextrose and dextrose syrups from starch
aspartase		conversion of fumarate to aspartate	amino acid production

(continued)

TABLE 8-9. (continued).

Principle Enzyme	Additional Enzymes	Primary Action on Substrate	Utilization
asparaginase		hydrolyzes asparagine	treatment of leukemia
B			
bromelain		hydrolyzes proteins	digestive aid
	protease, phosphatase, peroxidase	hydrolyzes proteins	meat, baking, pharmaceuticals, liquefying of gelatin desserts, preparation of fish concentrations of fish concentrates
C			
catalase		decomposition of H_2O_2 H_2O_2—O_2 + H_2O	food, brewing, textile, fur, rubber, leather, foam rubber, cosmetics, cheese, analytical, bleaching hair, cold sterilization of milk
cellulase		hydrolyzes cellulose and cellulose derivatives	citrus, essential oils and spices extraction, beer, pomace, pharmaceuticals, feed supplementation, vanilla extract, digestive aids, drain cleaning
	cellulase, pectin enzymes glucosidases	hydrolyzes cellulose, cellulose derivatives and related compounds	digestive aids, feeds, reduction of the viscosity of cellulose suspensions
	cellulase hemi cellulase	hydrolyzes cellulose and hemi cellulose	food and paper industry
carboxypeptidase			analytical
chymotrypsin			lens removal for cataract treatment, cells preparation and isolation, digestive aids
collagenase			preparation and isolation of cells

(continued)

TABLE 8-9. (continued).

Principle Enzyme	Additional Enzymes	Primary Action on Substrate	Utilization
D			
dextranase		hydrolyzes dextran	manufacture of glucose from corn syrup, dentistry (dissolve plaque), lower viscosity of syrup in sugar processing
diastase		hydrolyzes starch to glucose or dextrin	production of glucose from corn syrup
E			
elastase		hydrolyzes protein	rheumatoid arthritis treatment
acetylcholine esterase			cleaning of drains, analytical assay
F			
fumarase		convert fumarate to malate	fumarate production
ficin		hydrolyzes proteins	meat tenderizer, dissolve scrap film (in silver recovery from film)
G			
glucose isomerase		convert glucose to fructose	increase sweet content of products, high fructose corn syrup (HFCS) manufacture
α galactosidase		breaks down raffinose to galactose and sucrose	crystallization of sucrose from beet sugar molasses
galactose oxidase		convert galactose to galactonic acid	remove galactose from milk products
β-glucuronidase			medical diagnosis

(continued)

TABLE 8-9. (continued).

Principle Enzyme	Additional Enzymes	Primary Action on Substrate	Utilization
gelatinase	gelatinase, α-amylase, protease	solubilizes gelatin	photography
β-glucanase		degrade glucans	lower wort viscosity (beer making)
glucose oxidase		removes oxygen and glucose	fruit juice, food industry
		oxidizes glucose	analytical, research pharmaceuticals
	glucose oxidase, for use at low pH	removes oxygen	salad dressings
	glucose oxidase, low in catalase	oxidizes glucose	glucose test reagent
	glucose oxidase, catalase	oxidizes glucose	oxidase perosidase reagent
	glucose oxidase, catalase	removes sugar	eggs prior to drying
	glucose oxidase, catalase	removes oxygen from sealed containers	food packaging, salad dressing, sensitive food
	glucose oxidase, traces of cellulase, maltase invertase, amylase	removes sugar, oxygen	prevents off flavor in orange drinks, iron pickup in canned beer, rancidity of mayonnaise, enzymatic browning of frozen fruits; costabilization of ascorbic acid and vitamin B_{12}
	glucose oxidase, low in trace enzymes	removes oxygen	prevents off flavor in orange drinks
	catalase	oxidizes glucose to gluconic acid	
H			
histidinase		convert pyruvate to L-tryptophan	amino acid production

(continued)

TABLE 8-9. (continued).

Principle Enzyme	Additional Enzymes	Primary Action on Substrate	Utilization
hyaluronidase		hydrolyzes mucopoly-saccharide, hyaluronic acid	local anesthesia (sprains, bruises), insulin shock, preparation and isolation of cells, promotion of drug absorption
hemicellulase	hemicellulase, cellulase	hydrolyzes hemicelluloses and gums	antistalling agent, improve water holding capacity of rye flours, clean mucilage from coffee beans, break down oil well fracturing gels (gum)
I			
Invertase		hydrolyzes sucrose to glucose and fructose	invert sugar, soft and liquid centered candies, prevention of crystallization in jam
isoamylase		convert glycogen to poly-saccharides and oligo-saccharides	reduction of viscosity in beer
L			
lactase		hydrolyzes lactose to glucose and galactose	pretreatment of milk, whey treatment (production of glucose and galactose), prevent crystallization in ice cream
lipase		hydrolyzes fats to fatty acids and glycerol	improvement of flavor of ice cream, margarine, cheese, chocolate, cheesemaking (hydrolyzes milk fat), pharmaceuticals, digestive aids, improves whip-ping of egg whites
	lipase, protease	hydrolyzes fats	digestive aid formulation
	lipase, amylase trypsin	splits fat	pharmaceuticals, digestive aid, animal feeds, sewage disposal

(continued)

TABLE 8-9. (continued).

Principle Enzyme	Additional Enzymes	Primary Action on Substrate	Utilization
	lipase protease, amylase, pectinase	hydrolyzes fats and fatty acid esters cellulose	foods, pharmaceuticals
laccase			drying of lacquers
lipoxygenase			bleaching agent for white bread
lactate dehydrogenase			
lysozyme		hydrolyzes microorganisms	research, pharmaceuticals/opthalmic preparations, baby milk preparation
M			
malate dehydrogenase	lipase, protease, amylase	hydrolyzes fats	food brewing, textile, cosmetics, pharmaceuticals
N			
naringinase		convert naringin to prunin and rhamnose	debittering of juice
O			
Aromatic aliphatic oxidases		oxidizes aromatic & aliphatic carbons	conversion of oil to non-noxious residues
P			
peroxidase			
phosphomannanase		dissolves yeast cell walls	
pullalanase		convert starch to maltotriose	treatment of wort
proteases		hydrolyzes proteins	meat tenderizing, desizing textiles (silk and rayon fabrics), leather, meat tenderizer, pharmaceuticals,

(continued)

TABLE 8-9. (continued).

Principle Enzyme	Additional Enzymes	Primary Action on Substrate	Utilization
			protein hydrolyzates, desizing, degumming of silk, garment cleaning, feedstuffs for poultry, beef, cattle, swine; cereals, softening woolen fabrics, toothpaste, leather, chillproofing beer, meat tenderizing, digestive aids, formulations, dehairing hides
	protease, amylase	hydrolyzes proteins, starch	textile, paper, photographic products, laundry, brewing, processing acidulated fish, press water, baking, meat tenderizing
	protease, α-amylase, phosphatase	hydrolyzes proteins	leather
	protease, distase, lipase	hydrolyzes proteins, starch, fat	digestive aid
	protease, phosphatase, perosidase	hydrolyzes proteins	meat, brewing, photography, pharmaceuticals
	protease, amylase, gumase	hydrolyzes proteins, starch	feed ingredients for beef and dairy cattle, sheep, swine, poultry, chickens, turkeys
	protease, amylase, lipase	hydrolyzes starch, proteins, fat	digestive aid, meat tenderizing
	protease, cathepsin	hydrolyzes proteins, also milk curdling activity	digestive aid, pharmaceuticals
	protease, amylase, lipase, cellulase	hydrolyzes proteins, starches, fats, cellulose	mixture of enzymes from selected microorganisms and especially developed strains of bacteria to increase the efficiency of septic tanks, cesspools and dry wells

(continued)

TABLE 8-9. (continued).

Principle Enzyme	Additional Enzymes	Primary Action on Substrate	Utilization
	protease, amylase, lipase, cellulase	liquifies sewage	cleaning drain lines, grease traps, sewage treatment
	protease, amylase	hydrolyzes starch and proteins	meat tenderizing, pharmaceuticals, brewing
	alkaline protease	hydrolyzes proteins	detergents and cleaning compounds
	acid protease, amylase, lipase	hydrolyzes proteins, starches, fats in acid range	food, brewing, textile, cosmetic, pharmaceutical
papain		hydrolyzes proteins	meat tenderizer, prevents haze in chilled beer, softens high gluten flour (crackers), wound debride-ments, (medical), dentifrices, digestive aids, shrink-proofing of wool
	papain, chymopapain amylase, lipase pectase	hydrolyzes proteins	meat tenderizing, clarifying beer and malt beverages
pepsin		hydrolyzes proteins	animal feed additive, digestive aids, predigestion of cereals (precooked cereal), renin extender (curdling of milk)
penicillinase		destruction of penicillin by hydrolysis	pharmaceuticals, research, analytical
penicillin acylase		removal of side chains of penicillin	
polysaccharidase		hydrolyzes vegetable gums, mucilages	baking, sizing textiles, corn degermination, oil well drilling fluids
pectinase		hydrolyzes pectins	production of wines, production of concentrates, clarification of fruit juices

(continued)

TABLE 8-9. (continued).

Principle Enzyme	Additional Enzymes	Primary Action on Substrate	Utilization
pancreatin			leather bating, digestive aids
pullulanase		convert starch to maltose and maltotriose	improves carbohydrates utilization in brewing
R			
rennin		convert casein to paracasein	cheese manufacturing (curdling of milk), preparation of baby foods
S			
streptokinase		hydrolyzes proteins	anticoagulant therapy
T			
tyrosinase		convert tyrosine to dopa	medical (gout diagnosis, uric acid assay)
U			
uricase		convert uric acid to allantoin	
urease		decomposes urea to CO_2 and ammonia	removal of urea from blood, research, clinical tests in measuring urea in blood
V			
varidase			medical
X			
xantine oxidase		convert xantine to urate	xanthine assay

REFERENCES

1. Kirk and Othmer, *Encyclopedia of Chemical Technology,* 3rd Ed., Vol. 9, John Wiley & Sons, New York, NY (1983).
2. Atkinson, B. and F. Mavituna, *Biochemical Engineering and Biotechnology Handbook,* The Nature Press (1983); Oellerich, M., "Enzyme Immunoassays in Clinical Chemistry: Present Status and Trends," *J. Clin. Chem. Biochem., 18,* 197–208 (1980).
3. Savagerv, M. A., *Biochemical Systems Analysis,* Addison-Wesley Publishing Company, Reading, MA (1976).
4. Volkenstein, M. V., *Molecular Biophysics,* Academic Press, New York (1977).
5. Wolnak, B., "Status of U.S. Enzyme Industry 1978," in *Enzymes—The Interface Between Technology and Economics,* J. P. Danehy and B. Wolnak, eds., Marcel Dekker, Inc., New York (1980).
6. "Strong Growth Ahead for Industrial Enzymes," *C&EN,* 37–38 (June 19, 1981).
7. "Enzymes—Big Chance to Clean Up," *Chemical Week,* 89–90 (June 22, 1968).
8. "Enzymes Share of Laundry Market Grows, but Effectiveness of Product Is Disputed," *Wall Street Journal* (Oct. 9, 1969).
9. Free, A. H., "Status and Potential for Enzymes," in *Clinical Laboratory Practice in Enzymes,* J. P. Danehy and B. Wolnak, eds., Marcel Dekker, Inc., New York (1980).
10. Szabo, A. and E. Ors, "Enzymes Used in Clinical Chemical Analysis," *J. Clin. Chem. Clin. Biochem., 21,* 209–215 (1983).
11. Danielson, B., I. Lundstrom, K. Mosbach, and L. Stilbert, "A New Enzyme Transducer Combination: The Enzyme Transistor," *Anal. Letters, 12* (1311), 1183–1199 (1979).
12. Guilbault, G. G., "Ion-Selective Electrodes Applied to Enzyme Systems," *Ion Selective Rev., 4,* 187–231 (1982).
13. Guilbault, G. G., "Applications of Enzyme Electrodes in Analysis," *Ann. NY Acad. Sci., 369,* 285–294 (1981).
14. Guilbault, G. G., "Analytical Uses of Immobilized Enzymes," in *Enzyme Engineering,* Volume 6, E. Chibata, S. Fukui, and L. B. Wingard, eds., Plenum Publ. Co., New York (1982).
15. Guilbault, G. G., "Analytical Uses of Immobilized Enzymes," *Biotechnol. & Bioeng. Symp., 3,* 361–365 (1972).
16. Guilbault, G. G., "Applications of Enzyme Electrodes," in *Analysis, Biochemical Engineering II,* A. Constantinides, W. R. Vieth, and K. Venkatasubramanian, eds., Annals of New York Academy of Sciences, Vol. 369 (1981).
17. Kernevez, J. P., L. Konate, and J. L. Romette, "Determination of Substrate Concentrations by a Computerized Enzyme Electrode," *Biotechnology and Bioengineering, 25,* 845–855 (1983).
18. Mell, D. and J. T. Maloy, "Amperometric Response Enhancement of the Immobilized Glucose Oxidase Enzyme Electrode," *Analytical Chemistry, 48* (11), 1597–1601 (1976).
19. Mathis, D. E., F. S. Stover, and R. P. Buck, "Ion Transport in Free and Supported Nitrobenzene Aliquat Nitrate Liquid Membrane Ion-Selective Electrodes," *J. Membrane Science, 4,* 395–414 (1979).
20. Stover, F. S., T. R. Brumleve, and R. P. Buck, "Comparison of Time Constants for

Liquid Ion-Exchange Membrane Electrode Responses Determined by an Impedance Method and an Activity Step Method," *Analytica Chimica Act, 109,* 259–278 (1979).

21. Blaedel, W. M. and J. Wang, "Mixed Immobilized Enzyme-Porous Electrode Reactor," *Anal. Chem., 52,* 1426–1429 (1980).
22. Blaedel, W. J. and J. Wang, "Characteristics of a Rotated Porous Flow-Through Electrode," *Anal. Chem., 52,* 1697–1700 (1980).
23. Gough, D. A. and J. D. Andrade, "Enzyme Electrodes," *Science, 180,* 380–384 (1973).
24. Rawls, R. L., "Electrodes Hold Promise in Biomedical Uses," *C&EN,* 19–20 (Jan. 5, 1976).
25. Cooney, C. L., J. C. Weaver, S. R. Tannenbaum, D. V. Faller, A. Shields, and M. Jahnke, "The Thermal Enzyme Probe—A Novel Approach to Chemical Analysis," in *Enzyme Engineering, Vol. 2,* E. Kendall Pye and L. B. Wingard, eds., Plenum Press (1975).
26. Cooney, C. L., J. C. Weaver, S. R. Tannenbaum, D. V. Faller, A. Shields, and M. Jahnke, "The Thermal Enzyme Probe—A Novel Approach to Chemical Analysis," in *Enzyme Engineering, Vol. 2,* E. Kendall Pye and L. B. Wingard, eds., Plenum Press (1973).
27. Tran-Minh, C. and D. Vallin, "Enzyme-Bound Thermistor as an Enthalpimetric Sensor," *Analytical Chemistry, 50* (13), 1874–1878 (1978).
28. Mattiasson, B., B. Danielson, C. F. Mandenius, and F. Winquist, "Enzyme Thermistors for Process Control," *Ann. NY Acad. Sci., 369,* 295–305 (1981).
29. Mattiasson, B., B. Danielsson, C. F. Mandenious, and F. Winquist, *Enzyme Thermistors for Process Control, Biochemical Engineering II,* A. Constantinides, W. R. Vieth, K. Venkatasubramanian, eds., Annals of New York Academy of Sciences, Vol. 369 (1981).
30. Wisdom, G. B., "Enzyme Immunoassay," *Clin. Chem., 22* (8), 1243–1255 (1976).
31. Genbauer, C. R. and G. A. Rechnitz, "Immunoassay Studies Using Adenosine Deaminase Enzyme with Potentiometric Rate Measurement," *Analytical Letters, 14* (B2), 97–109 (1981).
32. Mattiasson, B. and H. Wilsson, "An Enzyme Immunoelectrode," *FEBS Letters, 78* (2), 251–254 (1977).
33. Gebauer, C. R. and G. A. Rechnitz, "Ion Selective Electrode Estimation of Avidin and Biotin Using a Lysozyme Label," *Anal. Biochem., 103,* 280–284 (1980).
34. Scharpe, S. L., W. M. Cooreman, W. J. Blomme, and G. M. Laekeman, "Quantitative Enzyme Immunoassay: Current Status," *Clin. Chem., 22* (6), 733–738 (1976).
35. Yuan, C.-L., S. S. Kuan, and G. G. Guilbault, "An Immobilized Immuno-Stirrer for the Determination of Creatine Kinase-MB Isoenzyme in Blood Serum," *Analytica Chimica Acta, 124,* 169–176 (1981).
36. Meyerhoff, M. E. and G. A. Rechnitz, "Electrode-Based Enzyme Immunoassays Using Urease Conjugates," *Anal. Biochem., 95,* 463–493 (1979).
37. Ngo, T. T. and C. Lenhoff, "Enzyme Modulators as Tools for the Development of Homogeneous Enzyme Immunoassays," *FEBS. Letters, 116* (2), 285–288 (1980).
38. Boitieux, J.-L., G. Desmet, and D. Thomas, "An 'Antibody Electrode,' Preliminary Report on a New Approach in Enzyme Immunoassay," *Clinical Chemistry, 25* (2), 318–321 (1979).

39. Janata, J., "An Immunoelectrode," *Jnl. Amer. Chem. Soc., 97* (10), 2914–2916 (1975).

40. Mattiasson, B. and H. Nilsson, "An Enzyme Immunoelectrode, Assay of Human Serum Albumin and Insulin," *FEBS Letters, 78* (2), 251–253 (1977).

41. Oellerich, M., "Enzyme Immunoassays in Clinical Chemistry: Present Status and Trends," *J. Clin. Chem. Clin. Biochem., 18,* 197–208 (1980).

42. Aizawa, M., S. Kato, and S. Suzuki, "Immunoresponsive Membrane: Membrane Potential Change Associated with an Immunochemical Reaction Between Membrane-Bound Antigen and Free Antibody," *J. Membrane Science, 2,* 125–132 (1977).

43. Aizawa, M., S. Kato, and S. Suzuki, "Immunoresponsive Membrane I. Membrane Potential Change Associated with an Immunochemical Reaction Between Membrane-Bound Antigen and Free Antibody," *Jnl. of Membrane Science, 2,* 125–132 (1977).

44. Aizawa, M., S. Suzuki, "An Immunosensor for Syphilis," *Jnl. of Solid-Phase Biochemistry, 4* (1), 25–31 (1979).

45. Kulys, J. J., "Development of New Analytical Systems Based on Biocatalysers," *Enzyme Microb. Technol., 3,* 344–352 (1981).

46. Al-Kaissi, E. and A. Mostratos, "Assessment of Substrates for Horseradish Peroxidase in Enzyme Immunoassay," *J. Immunological Methods, 578,* 127–132 (1983).

47. Guedson, J. L., C. Houanne, and S.Avrameas, "An Amplification System Using BSA—Antibody Conjugate for Sensitive Enzyme Immunoassay," *J. Immunological Methods, 57,* 133–142 (1983).

48. Mattiasson, B., H. Nilsson, and B. Olsson, "An Apoenzyme Electrode," *J. Applied Biochemistry, 1,* 377–384 (1979).

49. Pintruro, N. D., *Food Processing Enzymes, Recent Developments,* Noyes Data Corporation, Park Ridge, NJ (1979).

50. National Research Council, "The 1978 Enzyme Survey Summarized Data Prepared for Food and Drug Administration, PB81-216897 (1981).

51. Chang, T. M. S., "Use of Immobilized Enzymes in Medical Practice," in *Enzymes—The Interface Between Technology and Economics,* J. P. Danehy and B. Wolnak, eds., Marcel Dekker, Inc., New York (1980).

52. Chang, T. M. S., "Effects of Different Routes of *in vivo* Administration of Microencapsulated Enzymes," in *Enzyme Engineering,* Vol. 2, E. K. Pye and L. B. Wingard, eds., Plenum Press, New York (1975).

53. Hersh, L. S., "L. Asparaginase from *Escherichia coli II* and *Erwinia carotovura* Bound to Poly (methyl methacrylate)," in *Enzyme Engineering,* Vol. 2, E. K. Pye and L. B. Wingard, eds., Plenum Press, New York (1975).

54. Sherry, J., "Fibrinolytic Agent," *Disease of the Month,* Your Book Medical Publishers, Inc., Chicago (May 1969).

55. Terminello, L. and A. Lesuk, "Status and Potential for Enzymes as Parenteral Therapeutic Agents," in *Enzymes—The Interface Between Technology and Economics,* J. P. Danehy and B. Wolnak, eds., Marcel Dekker, Inc., New York (1980).

56. Anderson, J. L., H. W. Marshall, B. E. Bray, J. R. Lutz, P. R. Frederick, F. G. Yanowitz, F. L. Datz, S. C. Klausner, and A. D. Hagan, "A Randomized Trial of Intracoronary Streptokinase in the Treatment of Acute Myocardial Infarction," *New England Jnl. of Medicine, 308* (22), 1312–1318 (1983).

57. Khaja, F., J. A. Walton, Jr., J. F. Brymer, E. Lo, L. Osterberger, W. W. O'Neill, H. T. Colfer, R. Weiss, T. Lee, T. Kurian, A. Goldberg, B. Pitt, and S. Goldstein, "Intracoronary Fibrinolytic Therapy in Acute Myocardial Infarction," *New England Jnl. of Medicine, 308* (22), 1305–1311 (1983).

58. Venkatasubramanian, K., W. R. Vieth, and F. R. Bernath, "Use of Collagen Immobilized Enzymes in Blood Treatment," in *Enzyme Engineering, Vol. 2,* E. K. Pye and L. B. Wingard, eds., Plenum Press, New York (1975).

59. Broun, G., "Trends in the Use of Immobilized Enzymes and Proteins in Human Therapeutics," in *Enzyme Engineering, Vol. 2,* E. K. Pye and L. B. Wingard, eds., Plenum Press, New York (1975).

60. Rubin, D. H., "A Technology Assessment Methodology—Enzymes (Industrial)," The MITRE Corp., MTR6009, Vol. 4 (1971).

61. Atkinson, B. and F. Mavituna, *Biochemical Engineering and Biotechnology Handbook,* The Nature Press (1983).

62. Brezine, I. V. and N. F. Kazanskaya, "La Photographie Enzymatique," *La Recherche, 144,* 708–710 (1983).

63. Cheremisinoff, P. N. and R. P. Ouellette, eds. *Biotechnology, Applications and Research.* Technomic Publishing Co., Inc., Lancaster, PA; Basel, Switz. (1985).

9

Enzymes in Hostile Environments

INTRODUCTION

Nature is a vast laboratory in which experiments have been conducted over millions of years. Many blind avenues have been reached and the results discarded; a few successful efforts have endured. Countless variations have been tested, and opportunistic evolution has fed on this infinite variety. Part of this variety grows out of the multiplicity of environmental niches in which the process of evolution on this planet has taken place: deep seas; hot springs; caves; glaciers and snow fields; alkaline lakes; acid mines; sediments; volcanoes. Because existing and projected industrial environments reflect this multiplicity, it becomes important to study the behavior of organisms and their enzymes under extreme conditions.

This paper reviews the effects of salinity, temperature, light, xenochemicals, pH, oxygen, trace metals, and pressure on enzyme activity. Emphasis is on the marine environment because of its vastness (75% of the Earth's surface) and diversity (e.g., temperature, salinity), as well as its role in supporting life.

WATER AND SALINITY STRESS

Biota have been immersed in liquid media since primeval times. Nowhere is the celebration of time more obvious than in the marine environment.

The oceans and ancillary water bodies are a reservoir, a buffer, a growth medium. Their average temperature is only 3.8 °C, with a range from -2 °C to 30 °C. By comparison, the temperature range observed on continents goes from -83.3 °C to 105 °C in hot springs.

The waters of our world are a complex, variable, and interactive medium. The specific heat of water varies with its salinity and temperature. The density of sea water is directly related to temperature: for instance, at 5‰ salinity, the freezing point of water is -0.267 °C; at 40‰ salinity, it is -2.196 °C. The viscosity of water decreases with its temperature and increases slightly with its salinity. The surface tension of pure water decreases with increasing temperature. Conductivity, of biological importance in sea water, increases with temperature. Water is a covalent polar molecule; it has a large dipole moment, and a dielectric constant that decreases with increasing temperature. Dissolved oxygen decreases with temperature, at a rate that increases with higher chlorine content. The "average ocean" has a chlorinity of 19‰ and a salinity of 34.325‰ [1].

The archaebacteria *Halobacterium salinarium* lives under a 5 M salt concentration. The eubacteria *Klebsiella aerogenes* thrives at 3.4 M NaCl. Extreme halophilic bacteria have a cellular interior approaching 6 M KCl. A number of organisms are exposed to continual cyclic evaporative water losses and dessication often associated with dormancy or cyst production; under less drastic conditions, others are exposed to variation in the salinity of their internal milieu. The deleterious effects of KCl and NaCl on the activities of a number of enzymes have been documented in plants and animals in saline habitats.

Harsh cytoplasmic environments require specialized functional and structural proteins. The amino acid composition of halophilic proteins shows an increased usage of glutamic and aspartic acids relative to similar non-halophilic proteins. The analysis points to a highly hydrated halophilic protein surface as a method for maintaining structural integrity under high saline concentrations in a competition for water with cytoplasmic salts. These salt-tolerant bacteria are among the most primitive prokaryotes [2]. More recent organisms have opted for a genetically simpler strategy—mostly the acquisition of compatible solutes (such as amino acids, glycerol, and other polyols) to counteract the effect of ions [3,4].

TEMPERATURE

Heat is the energy of molecules in motion; temperature is the measure of the thermal state of a material. All biological systems need heat to drive their chemical reactions, to modify the materials they need, and to alter the properties of water.

Continued life in microorganisms is limited to the range of -10 °C to

70 °C. Plants and animals are limited to the range −2 °C to 40 °C. Cysts of microorganisms (bacteria, protozoa) have tolerated the lowest recorded temperature on earth (−83.3 °C), while live forms have been detected in hot springs at 105 °C [5]. Microorganisms are classified according to their temperature optimum as psychrophiles (optimum below 20 °C), mesophiles (optimum between 20 °C and 45 °C), and thermophiles (optimum above 45 °C).

Adaptations to temperature in whole organisms are the result of structural and functional adjustments. Some of these are genetically determined and are passed from generation to generation (adaptation); others are environmentally induced (acclimation, acclimatization) and involve gradual, although sometimes rapid, adjustments in individuals within genetic constraints [6,7]. Homeotherms maintain internal thermal homeostasis in spite of large variations in the temperature of their environment. They accomplish this through a number of regulatory mechanisms, including variation in metabolic heat production, vaso-distilation and vaso-constriction, perspiration, and insulation. Poikilotherms have an internal temperature that more or less follows the pattern of their immediate environment.

At the depth of submarine hydrothermal systems (2500–2650 meters), seawater remains liquid up to about 460 °C because of hydrostatic pressure. A variety of organisms, including chemosynthetic bacteria, are associated with such systems. Organisms living in superheated water use S, metals, and possibly organic gases and H_2S as energy sources. A recently isolated species from deep sea sulphide chimneys and warm water vents in the Pacific ocean, growing at temperatures in excess of 100 °C (but not below 70–75 °C), produced methane under laboratory conditions at 250 °C under a pressure of 265 atmospheres [8–12].

The native conformation of proteins is governed and determined by the sequence of amino acids in the primary structure; the folding pattern is thermodynamically controlled. The basic effect of elevated temperature is to supply enough kinetic energy to disrupt the chemical bonds. This effect is twofold: disruption of the bonds (e.g., hydrogen bonds) holding the secondary structure; and actual cleavage of the N-glycosinic and phosphate ester bonds. Heat denaturation is actually a cooperative phenomenon involving the entire protein-water complex [13–15].

Biological membranes are sensitive to temperature changes. A compensation in membrane fluidity occurs with temperature change over evolutionary times. This compensation is accompanied by differences in the saturation of phosphoglycerides. Observations on several proteins in many species confirm that protein thermostability is correlated with the microclimatic environmental conditions of the species [16,17].

Genetic adaptation is reflected in the functional and structural changes in parts (including enzymes) adapted to specific environments. Short-term adaptation does not involve substantial changes in amino acid sequences; rather, it involves more subtle mechanisms allegedly built into the enzyme structure [18–26]. Several examples of this adaptation can be given:

- A relationship has been shown between the heat resistance of collagen from various fish species and the amount of proline and oxyproline in the structure.
- The serum of some Antarctic fish freezes at -2 °C. This capability is associated with the presence of glycoproteins composed only of alanine, threonine, n-acetylgalactosamine, and galactose. The cis-hydroxyl groups of carbon 3 and 4 are responsible for this antifreeze property [27,28].
- Many proteins have conjugated carbohydrates on their primary structure. One of the roles of these carbohydrate moities is to increase the stability of the protein by protecting, among other things, against heat denaturation. The introdution of intrachain disulfide cross-links further stabilizes the protein (lysozyme), leading to a greater heat stability. When aldolase was treated with sulfydryl reagents, a new form evolved that is less resistant to heat but has enhanced activity [29,30].
- Aqueous solutions may be supercooled by several degrees centigrade, especially in capillary spaces. Bound water is resistant to freezing; hence marine organisms supercool and can withstand rather low temperatures. In a number of species, extracellular water is replaced by "antifreezes" that enable the organisms to withstand temperatures substantially below 0 °C. Certain insects may be supercooled to as low as -40 °C to -50 °C [27,31,32].

Most chemical reactions use heat as the source of the energy required to initiate the reaction, that is, to overcome the energy barrier to the reaction. Thermal activation is generally considered to involve the creation of a fraction of molecules having energies approximately equal to or greater than the activation energy, *E*. This can be expressed as

$$N/N_o = E^{-E/kT}$$

where N is the number of molecules with energy equal to or greater than E; N_o is the total number of reactant molecules present; k is the Boltzman constant; and T is the absolute temperature. The effect of temperature on a chemical reaction is then given by the empirically derived Arrhenius equation

$$\frac{d \ln k}{dT} = \frac{E}{RT^2}$$

Upon integration, we have

$$\ln k = \frac{-E}{R}\frac{1}{T} + \text{constant}$$

Most enzymes follow the Arrhenius plot over a range of physiological temperatures.

A fundamental thermodynamic equation is

$$\Delta F = \Delta H - T\,\Delta S$$

where ΔH is the heat of reaction, or the heat absorbed or released by a reaction; ΔF is the free energy or energy available for the reaction; and ΔS is the entropy or the number of possible configurations of a system having a given energy. At higher temperatures, the entropy of the system is increased, leading to a lower activation energy.

Adaptation to lower temperatures is equally interesting. Enzymes from organisms adapted to low temperatures are generally more active and efficient than those of organisms adapted to high temperatures, or not adapted at all. This increase in efficiency has been explained by greater flexibility in the enzyme structure; allowing favorable conformal changes. In controlled tests, such behavior has been shown by a variety of enzymes: acetylcolinesterase, ATPase, glyceraldehyde-3-phosphate dehydrogenase, and glycogen phosphorylase [33].

The free energy of the enzymatic reactions of poikilotherms is a few hundred calories/mole lower than this value for homeotherms. A linear relationship has been observed between ΔH and ΔS for a number of enzymes. The $\Delta H/\Delta S$ ratio of 330°K is similar to the ratio in reactions involved in breaking hydrogen bonds [33].

Hydrophobic bonds are exothermic and are stabilized at higher temperatures; hydrogen and electrostatic bonds are endothermic and are stabilized at lower temperatures. This is the basis for the hypothesis that attempts to explain the different bindings of enzyme and substrate as a function of temperature.

The adaptation of organisms (and their enzymes) to high and low temperatures offers a number of intriguing and potentially important properties.

Properties at High Temperatures

- Increased activity of enzyme, and hence an increased reaction rate
- Decreased viscosity of water
- Increased solubility of substrate and product
- Resistance to denaturation during extraction, leading to higher enzyme yield
- Increased shelf storage life
- Stability in the presence of detergent and organic solvents, observed in heat-adapted proteins
- Less sensitivity to denaturation in heat-adapted proteins
- Reduced possibility of contamination (sterilization effect)

Properties at Low Temperatures

- Significantly lower value of ΔG (activation energy)
- Protection of the substrate and products from degradation reactions
- No need for heat removal when operating at low temperatures, leading to cost savings

LIGHT

All life is ultimately derived from sunlight. The functions and behavior of whole organisms are greatly affected by light; the same should be true of enzymes.

The light reaching the earth from the sun changes in intensity and composition as a function of season, daytime cycle, altitude, latitude, cloudiness, atmospheric composition, and the presence of other masking materials (e.g., water and soil). Light is characterized according to its wavelength (λ), frequency (v), and energy (q) by simple relations:

$$v = c/\lambda$$
$$q = h\,v$$

where c is the speed of light (3×10^{10} cm/sec); and h is the Planck's constant (6.6236×10^{27} erg sec).

The most commonly observed effect of light on proteins and DNA is inactivation. However, light can be involved in more subtle functions. There are at least six pathways through which light can affect enzymes [34]:

- Direct activation or inactivation of an enzyme
- Indirect activation through the assistance of a light-sensitive effector

- Substrate activation of an inactive enzyme
- Modification of the reaction medium; such as pH
- *De novo* enzyme synthesis
- Modulation of enzyme activity

Each of these pathways has been clearly demonstrated in nature or in the laboratory.

The first step in any light activation reaction is the molecule's absorption of a quantum or radiant energy. As a result of this absorption, atoms and molecules become excited and can get rid of the extra energy by reemitting another quantum of radiation (fluorescence), by transferring the energy through collision to another particle, or by participating in a chemical reaction. Enzymes not sensitive to light can be made responsive by the addition of a non biological light-sensitive substrate or effector. Photochromic reagents have been applied to α-chymotrypsin, α-glucosidase, urease, β-amylase, α-amylase, and aldolase. Generally speaking, the result is a decrease in k_m (the quantity of substrate needed to achieve half the maximum speed of the reaction).

Slight irradiation of substrates enhances enzymatic reaction in some cases. However, the mechanism involved is not well understood. A number of enzymes are directly activated by light; examples include ornithine decarboxylase, citrate lyase ligase, bacterial hydrogenases, nitrogenase (in *Rhodopseudomonas capsulata*), and glutamine synthetase. Urocanase of *Pseudomonas putida* is modulated (deactivated-reactivated) by UV light. Obviously, carotenoids and chlorophylls demonstrate light-dependent steps in their biosynthesis [35].

A vast body of literature exists on the use of light to initiate or drive chemical reactions, although very few photochemical reactions have achieved industrial status. The known industrial photochemical synthetic processes include photochlorination, the preparation of caprolactam via the photochemical synthesis of the cyclohexane oxime, photochemical sulfonation, and photochemical hydrosulfidation. Several examples may be cited:

- Vitamin D is synthesized in human skin through the action of light. Most of the vitamin D production in the United States is based on UV synthesis. Both vitamin D_2 and D_3 can be synthesized in this manner: vitamin D_2 from ergosterol and vitamin D_3 from 7-dehydrocholesterol.
- The major recent development in photochemical technology is the polymerization and curing of polymers, coatings, inks, and related materials.
- The use of photochemistry for the preparation of photoresists in semiconductors production is a growing field.
- Photochemical techniques are also used in disinfection, sterilization,

control of biofouling, and pollutant removal. This involves the use of low-pressure UV lamps to disrupt the life-supporting mechanisms in a variety of microorganisms, or to cause chemical changes in pollutants that will facilitate their removal [36].

The effect of light in inducing or modifying biochemical reactions is virgin research territory. The action of light in driving photosynthetic apparatus and the cis-trans isomerisation of retinal (heart of the vision mechanism) are the best known examples. Bioluminescence, the active radiation of light, is a rather common phenomenon associated for most people with dancing fireflies and glowworms. It actually exists in almost all taxonomic groups. In deep sea below 400 meters, 20% of the organisms are bioluminescent. Luminescence exists in many bacteria and fungi. In most cases, the light emitting system involves the action of luciferase on luciferin in the presence of oxygen. As we look closer, the reaction becomes more complex.

Given the necessary research activities, it should be possible to use light as a control scheme to affect enzyme activity and productivity.

XENOCHEMICALS

More than once over the course of evolution, microorganisms have been presented with foreign chemicals as potential sources of carbon and energy [37]. With the advent of man-made chemicals, the stress of foreign substrates has become a daily occurrence. One of the very few benefits associated with the practice of discarding chemical waste in streams and on land is the evolution of microorganisms capable of metabolising new, large, and complex molecules. This ability arises from two sources: random mutations that alter a structural gene, giving rise to a new enzyme; or a modification in an existing enzyme regulatory mechanism, for instance, allowing a cell to synthesize an enzyme in the absence of its normal inducer [37].

It is obviously advantageous for microorganisms to have enzymes that are highly flexible in their ability to metabolize different substrates. The menu of potential substrates and the pathways available for metabolizing a single chemical seem limitless in soil microorganisms. Although specialization may exact a very high price, it provides speed of reaction, an attribute often favored by natural selection.

Soil pseudomonads owe their versatility to the breadth of the substrate specificity of their enzymes. Organisms of the *Pseudomonas* family are especially versatile. They have enzymes that allow the assimilation of a large number of hydrocarbons and aromatic molecules (metabolized by oxidoreduction) that are often very toxic. These organisms are found to function in a cooperative mode: a first organism modifies a molecule by assimilating it into its metabolism; it then rejects the molecule, which serves as a source of carbon for a second microorganism, and so on, until total degradation has been achieved. The secret of a microorganism's ability to attack a broad range of previously unknown chemicals, is its ability to catalyze the early steps in degradation, producing simpler molecules that can easily enter common metabolic pathways such as the Krebs cycle [37].

Modern chemistry deals with increasingly larger molecules that present new challenges to biodegradation. Highly chlorinated molecules such as many pesticides (e.g., DDT) and polychlorinated biphenyls (PCB) are a special challenge. Chakrabarty was the first to select artificially a mixed microbial culture able to completely metabolize 2,4,5 T. Note that this superbug was not created by recombinant DNA techniques, but by passing the appropriate plasmids (with the appropriate resistance gene) from one bacterium to another. The mutation and selection approach to discovering new microorganisms or well-known organisms with new capabilities may lack the glamor of recombinant DNA technology, however, it might bring to the laboratory an army of cells with new functions and capabilities.

pH EFFECT

Probably no stress is more damaging to life than a low pH. Bacteria, fungi, and eukariotic algae can live very close to a pH of zero; however, vertebrates survive in a narrow range, never below 3.5 to 4, while higher plants die below a pH of 2. A variety of intracellular chemical species, including deoxyribonucleic acid and adenosine triphosphate, are acid-labile. The B-glycosidol linkages in cell walls are sensitive to attack by acids.

Enzyme activity depends strongly on pH which has a number of important effects: it can affect ionogenic groups of the active site or of neighboring groups; it can change the substrates; it can modify co-enzymes; it can change the state of nonenzymatic components of the systems; and it can change the shape and structure of the enzyme locally or globally. Many laboratory studies have painstakingly documented these effects for a variety of specific enzyme-substrate combinations. We will not review those here, but will focus on a more general picture of the effect of pH on enzymes.

Using the Michaelis formulism, the mathematics of the pH effect is straightforward in simple cases of an acid system with two ionization states:

$$AH_2 \underset{K_1}{\rightleftharpoons} AH^- \underset{K_2}{\rightleftharpoons} A^{2-}$$

K_1 and K_2 are the two ionization constants:

$$K_1 = \frac{[AH^-][H^+]}{[AH_2]} \qquad K_2 = \frac{[A^{2-}][H^+]}{[AH^-]} = \frac{[A^{2-}][H^+]^2}{[AH_2]K_1}$$

$A_s = [AH_2] + [AH^-] + [A^{2-}]$ is the sum of all the concentrations of the three species. If AH^- is the only reactive species, the relation between the reaction rate, v, and the pH, $[H^+]$ is given by the equations

$$v = k\ [AH^-] \text{ and}$$

$$v = \frac{k\ A_s}{1 + ([H^+]/K_1) + (K_2/[H^+])}$$

This dependence has a maximum at

$$H^+ = (K_1\ K_2)^{1/2},$$

$$\text{or pH} = 1/2\ (p\ K_1 + p\ K_2)$$

This explains the well-known bell-shaped curve of the pH versus the reaction rate for enzymes.

OXYGEN TOXICITY

Oxygen is essential to life. Oxygen is toxic to life. How does this paradox arise?

The origin of life on earth is associated with an oxygen-free environment that lasted for over 2×10^9 years. The problem of oxygen toxicity to life arose with the evolution of oxygenic photosynthesis; the progressive invasion of the biosphere by oxygen produced by photosynthesis was one of the first cases of pollution on earth. Subtle mechanisms for dealing with this stress have evolved.

Oxygen itself is not a very reactive molecule and hence is not very toxic.

Some 95% of the available oxygen is utilized by the normal respiration pathway; that is, it is reduced to water by the cytochrome oxidase system. When oxygen is reduced gradually, two very reactive substances occur: hydrogen peroxide, H_2O_2, and the superoxide radical O_2^-. Fortunately, cells are equipped with enzymes that rapidly metabolize these toxic substances: catalases and peroxidases destroy hydrogen peroxide, while superoxide dismutases eliminate the superoxide radical. The superoxide dismutases have also been associated with protection against ionizing radiation, and have been implicated in the aging process through their accumulated failure to destroy all the scavenging radicals produced over the life of a cell. Iron and copper became associated with cells early as a method of protection against oxygen products. Dismutases and peroxidases contain these metals; most superoxide dismutases contain zinc (although it is not involved in the catalytic function), iron, and copper [38,39].

A prerequisite to the evolution of life in the presence of oxygen was the invention of enzymes able to metabolize (e.g., cytochrome oxidases) and carry (e.g., hemoglobins, hemocyanins, hemerythrins) oxygen to cells of multicellular organisms. However, a number of organisms can live without the oxygen so essential to most of life. Facultative aerobic bacteria such as *E. coli* can reproduce and prosper with or without oxygen. The presence of two dismutases allows them to adapt to increasing concentrations of oxygen in their immediate environment. Virtually all aerobic organisms contain both the dismutase and the catalase/peroxidase detoxification systems; strict anaerobics are generally deprived of both superoxide dismutase and catalase. The ability of *Oscilatoria limnetica* to shift from oxygenetic photosynthesis to anoxygenic photosynthesis allows this bacteria to survive under strenuous conditions. H_2 can serve in the photoassimilation of CO_2 in this cyanobacteria. Archaebacteria such as *Spirillum*, *Campylobacter*, *Thermoproteus*, and *Desulfuromanas* from hot sulfur laden springs can grow with carbon monoxide as their sole source of carbon provided that elemental sulfur is available as an electron acceptor. *Thermobacillus* and *Sulfolobus* used in mineral leaching oxides H_2S to sulfur and finally to sulfuric acid.

TRACE METALS

A number of metals are essential to life (iron, iodine, copper, manganese, zinc, cobalt, molybdenum, selenium, chromium, tin, vanadium, fluorine, silicon, and nickel); many are very toxic (e.g., mercury, lead, cadmium, and arsenic). Heavy metals such as lead and cadmium inhibit DNA biosynthesis. Exposure to arsenic, chromium, and nickel compounds is associated with incidences of specific types of cancer in man. Compounds of beryllium, cadmium, chromium, cobalt, lead, nickel, and zinc have been shown to induce cancer in experimental animals. Heavy metals are potential mutagenics.

A number of metals are added to the environment through the operation of cars, coal-fired power plants, smelters, and other activities. Many species, organs, tissues, and cells are known to accumulate and concentrate such metals, creating very concentrated microenvironments. *Caldariella*, a thermophilic-acidophilic bacterium (pH 2.6, temp. 60–70 °C), is known to oxidize copper sulfate minerals and molybdenates. It can tolerate a concentration of 2,000 ppm of hexavalen molybdenum in its immediate environment. A variety of bacteria, including *Pseudomonas*, *Alcaligens*, *Acinetobacter*, *Flavobacterium*, and *Aeronomas*, can methylate lead, mercury, and selenium to more volatile and more toxic species. Drainage from coal and mineral mining operations and mill tailings from mineral beneficiation operations are a fertile ground in which microorganisms generate a variety of intriguing biochemical reactions. Tailing assays record the presence of iron, copper, lead, zinc, gold, manganese, sulfur, arsenic, selenium, mercury, and cadmium. In such environments, metal sulfides are readily oxidized by oxygen and solubilized by water, and the metal ions further oxidized by bacteria.

About a third of the known enzymes have a metal in their active site or require metals for proper function. Metals have been shown to be enzyme activators: structure-promoter, substrate binder, Lewis acid, and electron carrier. A number of examples of enzyme/metal interaction can be cited [40]:

- The enzyme pyruvate decarboxylase exhibits activity only in the presence of magnesium. In fact, magnesium is required for the full activity of many enzymes: alkaline phosphatase, ATPase, hexokinase, and deoxyribonuclease (DNAase 1).
- Calcium is essential for muscle contraction.
- Liver alcohol dehydrogenase contains two functionally distinct Zn^{2+} ions. Carbonic anhydrase found in erythrocytes is potentiated by zinc. Carboxypeptidase is another enzyme requiring zinc.
- Vanadium is thought to be bound to a carrier protein in ascidian worms.
- The synergistic effect of chromium and insulin is well known, although the mechanism is unclear (inhibition of insulinase, binding of insulin to tissue, and stabilization of insulin have all been proposed). This metal also activates phosphoglucomutases.

- Mn^{2+} is tightly bound to pyruvate carboxylase and glutamine synthetase.
- Iron is involved in many oxygen carrier proteins—haemoglobin, myoglobin, erythrocruorin, and haemerythrin—and is essential for oxygen (cytochrome peroxidase) and oxygen products (catalase, peroxidase) handling. Iron is also part of transferrin, ferrodoxin, and rubredoxin.
- Cobalt is essential to the coenzyme vitamin B_{12}. This coenzyme works with glutamate mutase, dioldehydrase, and ethalonamine deaminase.
- A variety of copper-protein complexes function as oxidases using oxygen as an electron acceptor.
- Many reductases and oxidases require molybdenum cycling between +5 and +6 oxidation states. Reductases involved in the process of nitrogen fixing are of much interest to the biotechnologist.

The sequence of catalytic effectivity of metalloenzymes in many reactions is

$$Mn^{2+} \quad < Fe^{2+} \quad < Co^{2+} \quad < Ni^{2+} \quad < Cu^{2+} \quad < Zn^{2+}.$$

The conclusion is that enzymes and metals work hand in hand and cannot be separated. Some metal activators can be traded one for the other, opening some interesting possibilities for modifying enzyme-active sites.

PRESSURE

All organisms living below the surface of the sea are exposed to pressures in excess of 1 atmosphere; organisms have been detected on the sea floor at greater than 1,000 atmospheres. It has been discovered that increased pressures retard or accelerate a number of biological processes. The basic explanation is that augmentation of pressure tends to inhibit reactions in which molecular interaction requires a volume increase, and to stimulate a reaction in which a volume decrease is required. During an enzymatic reaction, a typical enzyme unfolds somewhat to accept the substrate (increasing its volume); therefore, the effect of pressure is generally deleterious. This was demonstrated with proteolytic enzymes, whose reactions are accelerated by subnormal pressures and suppressed by supranormal pressures. The effect of pressure is more than mechanical: pressure is known to modify the viscosity of water and the ionization state of various substances. High pressure is not found only in the oceans; soil bacteria have been recorded living under pressures of up to 1,000 atmospheres.

Without adaptation, biological activity is suppressed at pressures of 300 to 3,000 atmospheres at normal temperatures. Because of adaptation, several marine bacterial groups are resistant to substantial pressure changes.

Two reactions of interest are known to increase with pressure. First is bioluminescence enhancement (above the temperature optimum), which takes place because of an increase in the amount of active enzyme available. Second, working with malic dehydrogenase at high temperatures (105°), investigators have shown an increase in activity (up to 1,000 atmospheres); this is explained by the effect of pressure in protecting against heat denaturation [1].

CONCLUSION

Throughout evolution and continuing today, our planet has displayed a variety of ecological niches. Although much of the hydrosphere and lithosphere is inhospitable to life, the image that emerges from the preceding look at the natural and man-made environment is one of variety, adaptability, and the constant enrichment of well-established mechanisms used in the living world. Ecology is physiology under the worst possible conditions (claims an old motto). Indeed, when life is virtually untenable on earth, we find thriving microorganisms: from −12 °C (fungi, bacteria) to 105 °C (sulfide reducing bacteria); from a pH of 0 (*Thiobacillus thiooxidans*) to 13 (*Plectonema nistocorum*); from distilled water to a salinity of 6 M.

These achievements in clinging to adverse environments are far greater if one considers that many microorganisms withstand insults from several simultaneous stresses. Organisms are often faced with rapidly fluctuating environments and the simultaneous stresses of temperature, pressure, acidity, and low availability of nutrients. They have adapted by developing a multipotential metabolic pattern, and the ability to husband a wide variety of nutrients simultaneously. Salinity, temperature, light, xenochemicals, pH, oxygen, trace metals, and pressure all mold the function and behavior of organisms, cells, and enzymes in the natural environment. The possibility exists (and in some cases has been clearly demonstrated) that these agents can be used, alone or in combination, to activate, drive, control, direct, and terminate enzymatic reactions of interest.

The tradition of utilizing a benign, standard, bland medium to grow organisms and to conduct reactions, in the laboratory and in industry, is shortsighted. It uses only a small fraction of the arsenal at our disposal.

The curious scientists and the entrepreneurial industrialists would do well to study organisms under extreme conditions: to understand their adaptation mechanisms, to bring them to the laboratory, and to incorporate them in industrial reactions.

REFERENCES

1. Kinne, O. *Marine Ecology.* Wiley-Interscience (1972).
2. Yancey, P. H., M. E. Clark, S. C. Hand, R. D. Bowlus, and G. N. Somero, "Living with Water Stress: Evolution of Osmolyte Systems," *Science, 217,* 1214–1222 (1982).
3. Rao, J. K. M. and P. Argos, "Structural Stability of Halophilic Proteins," *Biochemistry, 20* (23), 6536–6543 (1981).
4. Shilo, M., "Strategies of Adaptation to Extreme Conditions in Aquatic Microorganisms," *Naturwissenschaften, 67,* 384–389 (1980).
5. Vallentyne, J. R., "Environmental Biophysics and Microbial Ubiquity," *Annals N.Y. Acad. Sciences, 108,* 342–352 (1963).
6. Ushakov, B. P., "The Problem of Associated Changes in Protein Thermostability During the Process of Speciation," *Helgolander Wissenschaftliche Meeresunters, 14,* 466–481 (1966).
7. Hazel, J. R. and C. L. Prosser, "Molecular Mechanisms of Temperature Compensation in Poikilotherms," *Physiological Reviews, 84* (3), 620–671 (1974).
8. Cavanaugh, C. M., S. L. Gardiner, M. L. Jones, H. W. Jannasch, J. B. Waterbury, "Prokaryotic Cells in the Hydrothermal Vent Tube Worm Riftia Pachyptila Jones: Possible Chemoautotrophic Symbionts," *Science, 213,* 340–342 (1981).
9. Fischer, F., W. Zillig, K. O. Stetter, and G. Schreiber, "Chemolithoautotrophic Metabolism of Anaerobic Extremely Thermophilic Archaebacteria," *Nature, 301* (10), 511-513 (1983).
10. Baross, J. A., M. D. Lilley, and L. I. Gordon, "Is the CH_4, H_2 and CO Venting from Submarine Hydrothermal Systems Produced by Thermophilic Bacteria?" *Nature, 298,* 366–368 (1982).
11. Baross, J. A. and J. W. Deming, "Growth of 'Black Smoker' Bacteria at Temperatures of at Least 250°C," *Nature, 303,* 423–426 (1983).
12. Brock, T. D., "High Temperature Systems," *An. Rev. Ecol. Syst., 1,* 191–220 (1970).
13. Low, P. S. and G. N. Somero, "Temperature Adaptation of Enzymes: A Proposed Molecular Basis for the Different Catalytic Efficiencies of Enzymes from Ectotherms and Endotherms," *Comp. Biochem. Physiol., 49B,* 307–312 (1974).
14. Ushakov, B. P., "Coupled Evolutionary Changes in Protein Thermostability," in *Molecular Mechanisms of Temperature Adaptation,* C. L. Prosser, ed., A.A.A.S., Wash., D.C. (1967).
15. Zale, S. E. and A. M. Klibanov, "On the Rule of Reversible Denaturation (Unfolding) in the Irreversible Thermal Inactivation of Enzymes," MIT Report, 6-25-83 (1983).

16. Hofer, R., H. Ladurner, A. Gattringer, and W. Wieser, "Relationship Between the Temperature Preferenda of Fishes, Amphibians and Reptiles, and the Substrate Affinities of Their Trypsins," *Comparative Physiology, 99,* 345–355 (1975).

17. Davis, H. C. and A. Calabrese, "Survival and Growth of Larvae of the European Oyster (*Ostrea Edulis L.*) at Different Temperatures," *Biol. Bull., 136,* 193–199 (1969).

18. Somero, G. N., "Temperature as a Selective Factor in Protein Evolution: The Adaptational Strategy of 'Compromise'," *J. Exp. Zool., 194,* 175–188.

19. Doig, A. R., Jr., *Stability of Enzymes from Thermophilic Microorgansism, Enzyme Engineering,* Vol. 2, E. K. Pye and L. B. Wingard, Jr., eds., Plenum Press, New York (1974).

20. Patton, J. S., J. C. Nevenzel, and A. A. Benson, "Specificity of Digestive Lipases in Hydrolysis of Wax Esters and Triglycerides Studies in Anchovy and Other Selected Fish," *Lipids, 10* (10), 575–583 (1975).

21. Ornston and W. K. Yeh, "Origins and Metabolic Diversity: Evolutionary Divergence by Sequence Repetition," *Proc. Natl. Acad. Sci., 76* (8), 3996–4000 (1979).

22. Wiggs, A. J., "Seasonal Changes in the Thyroid Proteinase of a Teleost Fish, The Burbot, *Lota lota* L.," *Can. J. Zool., 52,* 1071–1078 (1974).

23. Johnston, I. A., and G. Goldspink, "Thermodynamic Activation Parameters of Fish Myofibrillar ATPase Enzyme and Evolutionary Adaptations to Temperature," *Nature, 257,* 620–622 (1975).

24. Rossebo, L. and B. Underdal, "Studies in Proteolytic Enzymes and Hemolysins from Pyloric Caeca of Cod, Herring and Mackerel Resistance to Heat and Gamma Radiation," *Archiv fur Fischereiwissenschaft, 23* (2), 128–135 (1972).

25. Johnston, I. A. and N. J. Walesby, "Molecular Mechanisms of Temperature Adaptation in Fish Myofibrillar Adenosine Triphosphates," *Comparative Physiology, 119,* 195–206 (1977).

26. Slott, S., G. Madsen, and B. E. Norman, "Application of a Heat Stable Bacterial Amylase in the Starch Industry," in *Enzyme Engineering, Vol. 2,* E. K. Pye and L. B. Wingard, Jr., eds., Plenum Press, New York (1974).

27. DeVries, A. L., "Glycoproteins as Biological Antifreeze Agents in Antarctic Fishes," *Science 72,* 1152–1155 (1971).

28. Lin, Y. and J. K. Gross, "Molecular Cloning and Characterization of Winter Flounder Antifreeze cDNA," *Proc. Natl. Acad. Sci., 78,* 2825–2829 (1981).

29. Thompson, T. L., W. E. Militzer, and C. E. Georgi, "Partial Denaturation of a Bacterial Aldolase Without Loss of Activity," *J. Bact., 76,* 337–341 (1958).

30. Johnson, R. E., P. Adams, and J. A. Rupley, "Thermodynamics of Protein Cross-Links," *Biochemistry, 17* (8), 1479–1484 (1978).

31. Meyer, G. H., M. B. Morrow, and O. Wyss, "Antarctica: The Microbiology of an Unfrozen Saline Pond," *Science, 138,* 1103–1104 (1962).

32. Marshall, J. J. and M. L. Rabinowitz, "Enzyme Stabilization by Covalent Attachment of Carbohydrate," *Archives of Biochemistry and Biophysics, 167,* 777–779 (1975).

33. Hultin, H. O., "Enzymes from Organisms Acclimated to Low Temperatures," in *Enzymes,* J. P. Danehy and B. Wolnak, eds., Marcel Dekker, Inc., New York, NY (1980).

34. Hug, D. H., *Photoactivation of Enzymes; Photochemical and Photobiological Reviews,* Vol. 6, K. C. Smith, ed., Plenum Press, New York, NY (1981).

35. Berezin, I. V., *Initiation of Enzyme Reactions by Light; Enzyme Engineering,* Vol. 2, E. K. Pye and L. B. Wingard, Jr., eds., Plenum Press, New York (1974).

36. Roberts, R., R. P. Ouellette, M. M. Muradaz, R. F. Cozzens, and P. N. Cheremisinoff, *Applications of Photochemistry.* Technomic Pub. Co., Inc., Lancaster, PA (1984).

37. Degradation of Synthetic Organic Molecules in the Biosphere, Proceedings of a Conference, San Francisco, CA, June 12–13, 1971, National Academy of Sciences Washington, D.C.

38. Fridovich, I., "Les Superoxydes ou les Dangers de la Vie Aerobie," *La Recherche, 9* (92), 743-749 (1978).

39. Frieden, E., "Biochemical Evolution of Iron and Copper Proteins," *C&EN,* 42–46 (March 25, 1974).

40. Phipps, D. A. *Metals and Metabolism.* Clarendon Press, Oxford (1976).

10

Transducers in Biotechnology

INTRODUCTION

Fermentation and tissue and cell culture are the essential bases for all biotechnology. To understand the phenomena involved, control the reaction rates, sense the health of the organisms; assaying, testing, measuring, monitoring, are required.

The first step in understanding, controlling and optimizing a process is the precise, accurate, and timely (ideally continuous) monitoring of important parameters. The state-of-the-art in automated monitoring is very advanced in mature industrial sectors. Biotechnology is a new field of endeavor and is struggling with the problems of adapting sensors developed for other applications or designing new sensors to satisfy its needs.

MEASUREMENT SYSTEM

From a systems point of view, a measurement is achieved through the use of a meter or sensor expanding the human senses ability to detect, measure and quantify [1]. Such transducers, basically convert one form of energy to another. For instance, a pressure measurement can be obtained by a piezoelectric crystal converting pressure (mechanical energy) to electrical current (electric energy). The following matrix (Table 10-1) identifies a few of these transducers, most of which have been developed to a high level of sophistication and performance. Because of the availability of microcomputers and their ability to interface directly

TABLE 10-1. Transducers as Energy Conversion Devices.

To \ From	Thermal	Mechanical	Chemical	Radiant	Nuclear	Electrical Magnetic
Thermal		Joule-Thompson Effect	Chemical Reactions Enzymes		Nuclear Reactor	Peltier Effect
Mechanical	Thermometer Vibrating quartz crystal	Scale Ruler Strain gage	Muscle	Radiation pressure		
Chemical	Endothermic Chemical Reactions		Chemical reactions Enzymes	Photosynthesis		
Nuclear						
Radiant	Thermoluminescence Optical fiber	Triboluminescence Optical fiber	Fluorometer Bioluminescence Enzyme Colorimeter	Optical fiber	Photomultiplier tube Radiation badge	Light bulb Optical fiber
Electrical Magnetic	Thermistor Pyroelectric Thermocouple Junction diodes Transistors	Piezoelectric Magnetic flow meter Impedance flow meter Electropneumatic Quartz crystal	Electrodes NMR Battery Mass spectrometer IsFET/ChemFET	Solar cell Optical sensor	X-ray detector	EKG monitor EEG monitor

with measuring instruments, our interest is mostly restricted to devices with electrical output.

Often it is not possible to measure an interesting property directly; a surrogate measure is then used (Figure 10-1). An example is measuring viscosity as an indicator of cell mass; another example is the use of culture optical density as a measure of cell density.

A systems view of measurement requirements will document the difficulties associated with taking the pulse of a complex medium such as a fermentation broth. An ideal meter should measure a most basic property, ignoring or correcting for interferences. The ideal sensor will have a number of characteristic requirements varying from one case to the other.

Accuracy

Accuracy is a measure of how close the empirical measurement is to the true value; or its conformity to an accepted standard value.

Sensitivity

Sensitivity is the lowest limit of detection required. pH information might be accurate enough in tenths of units but temperature detection must be as low as 10^{-6} °C for microcalorimetry.

Specificity

A fermentation broth is a diverse heterogeneous broth. The sensor must detect only the condition or molecules of interest. Selective membranes play a major role in targeted measurements.

Output Information

Ideally the output signal should be electrical or at least should not require human interpretation.

Independence from Interferences

An ideal sensor is immune to interfering substances or conditions (temperature, pH) or can easily compensate and correct for such interferences.

The impact of interfering substances can be minimized by selectivity or compensated for by different measurements or resolved mathematically by solving simultaneous equations.

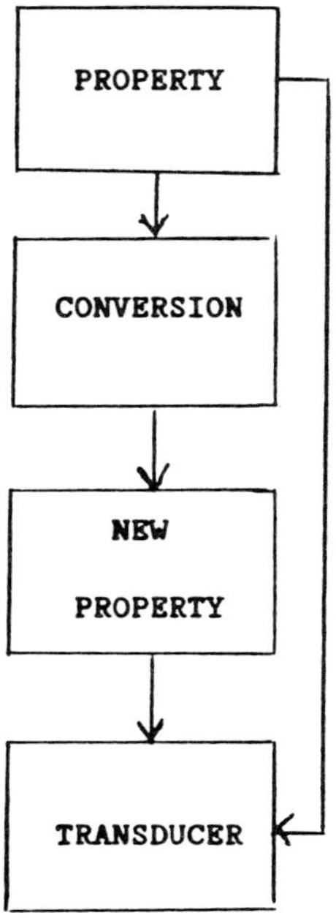

Figure 10-1. Measurement pathways.

Reproducibility

Measurements made under standard conditions should be repeatable from day-to-day and from laboratory to laboratory.

Stability

Sensors behavior must be stable enough to provide a meaningful measure. Stable instruments minimize the need for frequent calibration against standards.

Response Time

The response time of a sensor must be in keeping with the speed of change in the process and the intended use for the information.

Sterilization

Devices used in biotechnology must ideally be steam sterilized, or sterilized by chemical or radiation means.

Lack of Toxicity/Contamination

The material used to construct the sensing device must not be toxic to the microorganisms. The act of sampling, sensing and analyzing should not introduce contamination in a biological culture.

Environment

Sensors used in a biotechnology environment must be rugged and withstand attack from a variety of chemicals species besides temperature, acidity, salinity, and water.

Size

Sensors must be small enough to be introduced at a variety of positions within a medium, and must require a small enough sample for their operation. In other words, they must be unobtrusive and must not disturb precarious equilibrium conditions.

There are many ways to measure a property. Let's look at temperature as an example (Figure 10-2) [2,3]. Temperature can be measured with a thermometer functioning on the principle of an expanding liquid as a function of temperature. Range and precision are dependent on liquid selection and capillary size. A resistance thermometer is a resistor which is temperature sensitive. It produces a voltage when a current is passed through it. A thermistor is a semiconductor which exhibits a large change in resistance for a relatively small change in temperature. Thermopiles are a combination of thermistors. Diodes are temperature sensitive and thus presently can be used to measure temperature [4]. Metal oxide field effect transistors, as well as ceramic materials, respond to temperature [5]. Calorimeters are devices used to measure heat evolution with great precision [6,7]. Certain phosphors emit radiation at specific wavelengths as a function of temperature [8,9]. This light information can be collected and transmitted through an optical fiber [10].

This discussion points out the richness of our technology and the inherent difficulty in making an appropriate choice of measuring system for a specific application.

THE STATE OF THE ART

Process parameters such as power-to-shaft, agitator speed, cooling jacket flow rate and temperature belong to the world of chemical

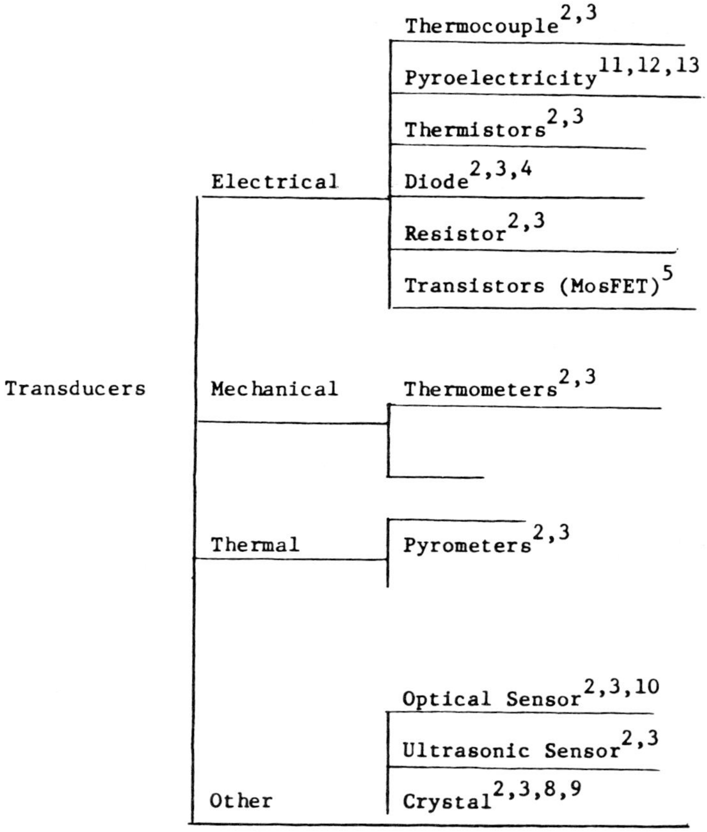

Figure 10-2. Temperature measurements.

engineering and their measurement is part of the routine practice of engineering based on well established theories and practical knowledge.

A number of sophisticated instruments have appeared in the laboratory: liquid chromatographs, gas chromatographs, mass spectrometers, ultraviolet/visible/infrared spectrometers. These devices are sensitive and can be adapted to measure a wide spectrum of species. Unfortunately, most of these have not been designed with the industrial environment in mind.

The technology to measure physical conditions such as temperature, flow, pressure, viscosity, is well established for batch and continuous operations.

- Viscosity is a measure of the resistance of a fluid to flow. Many viscosimeters for batch and on-line measurements based on capillary tube, rotating cylinder and oscillating disk methods are on the market [14].

- Pressures are best measured with piezoelectric pressure transducer with integrated temperature compensation or with pressure sensitive transistors [15,18].
- pH is a measure of the concentration of hydrogen ions a_H+

$$pH = -\lg a_H+$$

The electrochemical measure of hydrogen ion concentration follows the Nernst equation.

$$E = E^\circ + kT\, pH$$

where E is the electrode potential, E° the reference electrode, and T the absolute temperature.

- The redox potential measures the state of oxydo reduction reactions. The half reaction applicable to an indicator redox couple is

$$In_{ox} + ne^- \rightleftharpoons In_{red}$$

at a potential E, the ratio of concentrations of the two forms follow the Nernst equation

$$R = E^\circ_{In} - \frac{RT}{nF} \ln \frac{[In_{red}]}{[In_{ox}]}$$

- Gas flow rates are often measured with variable area flowmeters. Liquid flow rates can be measured with electromagnetic flowmeters. Exit gas can readily be measured: O_2 by paramagnetic analyzers and CO_2 by thermal conductivity or long path infrared analyzers [19].

Biomass, population density, cell counts are all of interest and provide complementary information toward assessing the quantity and health status of the cellular organisms. Most of the methods in use are indirect measures. A direct cell count can be obtained by a dilution, plating, growing, counting cycle. A more real time method involves the use of a coulter counter to provide a cell count as well as a size distribution curve. Many parameters are intimately correlated. Biomass concentration is often estimated from turbidity, gas balance, heat evolution, or fluorescence. Laser flow microphotometry allows to count cell, sorts them according to size, discriminates between species and even gives information on the internal composition of cells [20–22]. Bacterial count has also been obtained by reacting luceferase with adenosine triphosphate or measuring the conductivity of a solution. Let's review the state of affairs.

- Cell mass concentration during fermentation processes is an essential but difficult measurement. One approach is to measure the viscosity of the cell broth. In general, the viscosity of a fermentation broth depends on a number of factors: biomass concentration, biomass morphology, extracellular products concentration, and temperature. These parameters must be controlled to the extent that variations in these do not cause significant changes in broth viscosity [14,23].
- Several devices for continuously monitoring the optical density (OD) of microorganisms in cultures are commercially available. Using such a device, a side stream from a fermentation culture can be routed through a spectrophotometer and returned to the culture broth. A disadvantage in using these devices is that only relatively low cell densities can be determined. For measuring high cell density, dilution of the cell suspension is the only choice. During continuous monitoring of concentrated cell cultures, an on-line dilution scheme would require both an accurate dilution device and a continual draw-off of a waste stream. Both of these options are impractical and expensive [24].

  In an alternative method a modified flow cell is placed in the spectrophotometer and culture broth is circulated by a circulating pump through the annular space, bottom to top, and then returned to the fermenter. Different dilution devices with varying outer diameters of the inner tubes can be used to measure the cell density of a given culture. This device offers a dilution of the culture stream without physical dilution of the culture broth.

  In another device, a submersible probe permits accurate, easy, and continuous measurements of the optical density (OD) of a fermentation solution by removing air bubbles which interfere. In the measuring chamber, the light emitted from the light source passes through the broth and is projected on a photocell or a phototransistor. The optical density of the broth can be determined from the photoelectromotive force given by a DC-voltmeter connected to a photocell or a phototransistor [25].
- Turbidity can be a useful means of monitoring cell growth in a fermentation vessel. To avoid the time and labor involved in measuring turbidity from samples taken from a fermenter, an in-tank measurement device has been developed. This device is based on a differential measurement from a single light source [26].

- Calorimetry is a suitable method for monitoring changes in microbial energy transformation. During microbial growth processes, a large amount of heat is generated. It has been shown experimentally that the heat accompanying growth of microorganisms is closely related to the catabolic activity of the cells. The calorimetry technique has the advantage of being suited for measurement of heat evolved during anaerobic as well as aerobic growth processes. Calorimetry also allows a check on material balances during fermentation [6,27].
- The Coulter counter is an electronic device for measuring microbial cell number and size distribution. The device consists of a reservoir containing an electrolyte, two electrodes connected to an ammeter, and a tube with a small orifice to which a vacuum is applied. A cell culture is placed in the reservoir and an electrical potential is applied; as cells pass through this orifice, the electrical resistance increases causing detectable pulses in the current. By counting these pulses, an estimate of cell number is obtained. By analyzing the height of the pulses, an estimate of cell size distribution can also be obtained [28].
- A conductivity apparatus for measuring aerobic and anaerobic activity is comprised of two electrode systems, each having a cathode and an electrolyte inside, a liquid junction for contact with an outside liquid, and an exposed anode. In one of the electrode systems, the exposed anode is covered with a microorganism-impermeable membrane. The microorganism activity during the fermentation is equal to the difference in current value between the two electrode systems, namely the microorganism and the liquid medium.
- Cell concentration is an important state variable in many fermentations. Continuous, accurate measurements of cell concentration is essential to the development of maximum yield during continuous fermentations. The relationship between impedance changes of microbial cultures and cell concentration has been described quantitatively in laboratory studies [29].

The difficulty, for sure, rests with measuring intermediate metabolites in the complex broth. Microspectrophotometry and microfluorometry have been used to measure small quantities of a variety of substrates. Immobilized whole cells have been used as BOD sensors in the form of a microbial electrode [30]. Immobilized enzymes are increasingly being used in automated analyses [31].

Ion-selective electrodes are increasingly used to measure pH, gas, ions, enzymes and organic substrates. They are small, rugged and accurate [32,33].

The response of an ion selective electrode to an ion A of activity a and charge Z in the presence of an interfering ion B is given by

$$E = \text{const.} + \frac{2.303\ RT}{Z_A F}\ \lg\ [a_A + K_{A,B}^{pot}\ (a_B)^{Z_A/Z_B}]$$

Where E is the measured potential, R is the gas constant (8.314 JK^{-1}), T is the temperature (°K), F is the faraday constant (96,487 °C Mol^{-1}) and $K_{A,B}^{pot}$ is the selectivity constant.

There are several types of ions selective electrodes.

Glass Electrodes

The composition of the glass determines the selectivity. Glass electrodes are available to measure pH and cations concentrations.

Membrane Electrodes

A selective membrane permeable to ions or gases which permeates the membrane and alters the pH of an internal filling solution can be made selective to very specific species in the presence of interferences.

Enzyme Electrodes

The electrode can be covered with a membrane supporting an immobilized enzyme. The electrode detects a product of the enzyme/substrate reaction which can be correlated with substrate concentration. Glucose oxidase so immobilized will measure glucose concentration through a pH change, the disappearance of O_2 on the production of oxygen.

Chemical field effect transistors operate on the principle of current flow upon charge induction into the semiconductor responding to a charge particle on its surface. Metal oxides semiconductors have been used to measure pH. Ion selective field effect transistors in combination with selective membranes have been applied to measure a variety of ions. Chemical field effect transistors are being developed to measure a variety of organic substrates by measuring the product of a specific enzymatic reaction.

Their small size, ruggedness, ability to be introduced virtually anywhere in a fermentation process, combined with their specificity, sensitivity and electrical output will insure an important future role for them in fermentation monitoring and process control.

Enzyme thermistors [34,35] are a new addition to the arsenal. The basic idea is to measure the heat evolved during exothermic enzymatic reaction. The heat produced can be correlated with the activity of the enzyme and hence with the substrate concentration.

The temperature due to a chemical process can be measured by calorimetry. A knowledge of the moles of product formed (n) and the heat capacity of this system (k), yields ΔH.

$$\Delta H = \Delta T(k)/n$$

Optical fiber-based sensing techniques including: reflective (on-off switching); intensity (amount of light) for measuring temperature; microbending (squeezing the fiber varies light amplitude) for measuring pressure; interferometric (phase differences); and polarimetric (polarization of light); are appearing on the market and are likely to cause a revolution of sort [10].

Most sensors used in biotechnology today measure average bulk medium characteristics. Ultimately we are interested in the intracellular state of affairs. A number of nondisruptive techniques are under development to measure intracellular components. Fluorescence emission spectroscopy can determine nicotinamide adenosine nucleotide (NADH) inside a cell. The rate of NADH to NAD^+ conversion is a good indicator of oxido-reduction reactions.

Figure 10-3 is an indication of the availability of sensors to measure parameters of interest to the biotechnologist.

INTERPRETATION

Measurement is an active process and the possibility of disturbing a precarious balance and measuring the disturbances created is always present when we deal with fragile and capricious biological systems.

Interpretation of findings is the most difficult part of measurement. The conditions of the measurement and the basic phenomena involved must be fully understood in order to extract meaning from physical measures. Any measurement implies a model of a deeper phenomena.

Often measurements are like images or light reflected in a mirror. Few measurements are made at the source—the atomic and molecular level. The more removed we are from sampling the phenomena, the more arduous is the explanation afforded by the data.

Most measurements are measures of central tendency in a continuously changing population. The statistical nature of the data should not be forgotten.

In a continuously changing environment static measures are of little value. This is the case for most biological phenomena. Because of the stochastic nature, an average value without an error band, limits the ability to properly interpret the findings.

PARAMETERS	EXAMPLES OF METER	AVAILABILITY Laboratory	Plant
Cellular Material			
Count	Coulter counter Laser flow cytophotometry	■	■
Volume			
Mass			
Physical Parameters			
pH	Electrode IsFET	■	■
Temperature	Thermistor Microcalorimetry	■	■
Flow	Magnetic flow meter	■	■
Pressure	Piezoelectric	■	■
Dissolved gases			
O_2	Electrode/Polarography	■	■
CO_2		■	
Redox Potential		■	
Exit Gases			
O_2	Paramagnetic	■	■
CO_2	Infrared	■	■
Heat Evolution		■	■
Chemical Species			
Ions	Enzyme electrode ChemFET	▼	
Organic Substrates		▼	
Enzyme/Protein	Enzyme electrode	▼	
Cofactors	Spectroscopy	▼	

▼ Available only for some species.

Figure 10-3. State of the art.

CONCLUSIONS

Several trends in instrumentation design can be detected.

Miniaturization

Miniaturization is the strongest stimuli for the phenomenal growth in microelectronics. The same energy is now being applied to minaturizing the bulky mechanical components of instruments. The engineers are designing very small devices. A recently announced portable gas chromatograph is laid on a 5 × 5 cm silicon wafer [36–39].

Electrical Output

The form of the output signal from an instrument and its compatibility with downstream processing is rapidly becoming a major concern. The world has made the transition to digital computers, hence devices with electrical output are favored.

On-Board Computing

Instruments are increasingly being endowed with memory and computation power. The ability to perform computation, compare the results to a standard or store information for known chemical signatures is becoming the norm [40–42].

Multiple Parameter Measurements

Single measurements provide limited information. The value of measurement correlation is truly synergistic [43].

Dynamic Measurements

Biological reactions are relatively rapid. The need to capture vanishing species and the interest in elucidating dynamic reaction mechanisms, and probing temporal phenomena have given rise to micro and picosecond measuring systems based on short pulses of laser light [44].

Optical Sensing

Optical sensing by the watchful eye of the observer or by colorimetry was for a long time the major method of measuring changes in chemical reactions and biological processes. Progress in optical fibers as sensing devices is injecting new life in this well established area. Optical sensors measure environmental changes by looking at how light is absorbed, transmitted, reflected or emitted. Sensors based on fiber optics are becoming available to measure rotation, temperature, electric current, sound, and liquid levels [10].

The most dramatic progress is made at the interface between traditional discipline. Instrument design to satisfy the needs of biotechnology forces the integration of physics, chemistry, biology, engineering and computer science. This synergism will give rise to a new family of instrumentations to the benefit of each discipline.

REFERENCES

1. Weaver, J. C. and S. K. Burns, "Potential Impacts of Physics and Electronics on Enzyme-Based Analysis," *Applied Biochemistry and Bioengineering, 3,* 271–308 (1981).
2. Kennedy, R. H., "Selecting Temperature Sensors," *Chemical Engineering, 8,* 54–71 (Aug. 1983).
3. Erb, G., "Specifying Temperature Measuring Systems," *Test & Measurement World,* 32–43 (June 1983).
4. Barth, P. W. and J. B. Angell, "Thin Linear Thermometer Arrays for Use in Localized Cancer Hyperthermia," *IEEE Trans. on Electron Devices, 29* (1), 144–150 (1982).
5. Nitta, T. Jiroterada, and F. Fukushima, "Multifunctional Ceramic Sensors: Humidity-Gas Sensor and Temperature-Humidity Sensor," *IEEE Transaction on Electron Devices, 29* (1), 95–101 (1982).
6. Erikson, R. and T. Holme, "The Use of Microcalorimetry as an Analytical Tool for Microbial Processes," *Advances in Microbial Engineering,* B. Sikyta, A. Prokop, and M. Novak, eds., Biotechnology and Bioengineering Symposium No. 4, 581–590, John Wiley and Sons, New York, NY (1973).
7. "Calorimeter Aids Process Optimization of Fermentation," *Biotechnology News, 3* (14), 3.
8. Oliver, D. G., A. H. Sanders, L. Jang, D. Poy,and A. Van Heuvelen, "Liquid Crystal Thermography. A Method for Monitoring Temperature Gradients in Microtitration Plates," *J. Imm. Methods, 58,* 25–29 (1983).
9. Sharpless, E. N., "Liquid Crystals for Temperature Measurement," *Test & Measurement World,* 26–28 (June 1983).
10. Hecht, J., "Fiber Optics Turns to Sensing," *High Technology,* 49–56 (July-August 1983).
11. Patel, J. S. and D. M. Hanson, "Pyroelectric Luminescence," *Nature, 293,* 445–447 (1981).
12. Rahnamai, H., "Pyroelectric Differential Thermal Analyzer," *Ferroelectrics, 38,* 963–966 (1981).
13. Rahnamai, H., "Pyroelectric Enthalpimetric Sensor," *Anal. Chem., 54,* 142–143 (1982).
14. Perley, C. R., J. R. Swartz, and C. G. Cooney, "Measurement of Cell Mass Concentration with Continuous-Flow Viscometer," *Biotechnology and Bioengineering, 21,* 519–523 (1979).
15. Nye, J. F. *Physical Properties of Crystals.* Clarendon Press, Oxford (1967).
16. Kepler, R. G., "Piezoelectricity, Pyroelectricity and Ferroelectricity in Organic Materials," *Ann. Rev. Phys. Chem., 29,* 497–518 (1978).
17. Kolm, E. and H. Kolm, "How Piezoelectric Devices Earn Their Keep," *ChemTech,* 180–186 (March 1983).
18. Lee, Y. and K. D. Wise, "A Batch-Fabricated Silicon Capacitive Pressure Transducer

with Low Temperature Sensitivity," *IEEE Trans. on Electron Devices, 29* (1), 42–46 (1982).

19. Shoda, M. and Ishikawa, Y., "Carbon Dioxide Sensor for Fermentation Systems," *Biotechnology and Bioengineering, 23,* 461–466 (1981).
20. Gregg, E. C. and K. D. Steidley, "Electrical Counting and Sizing of Mammalian Cells in Suspension," *Journal of Biophysics, 5,* 393 (1965).
21. Olson, R. J., S. L. Frankel, and S. W. Chisholm, "An Inexpensive Flow Cytometer for the Analysis of Fluorescence Signals in Phytoplankton: Chlorophyll and DNA Distributions," MIT Report 5-43-83 (1983).
22. Galbraith, D. W., K. R. Harkins, J. M. Maddux, N. M. Ayres, D. P. Sharma, and E. Firouzabady, "Rapid Flow Cytometric Analysis of the Cell Cycle in Intact Plant Tissues," *Science, 220,* 1049–1051 (1983).
23. Perley, C. R., J. R. Swartz, and C. L. Cooney, "Measurement of Cell Mass Concentration with a Continuous-Flow Viscometer," *Biotechnology and Bioengineering, 20,* 519–523 (1979).
24. Lee, C. and H. Lim, "New Device for Continuously Monitoring the Optical Density of Concentrated Microbial Cultures," *Biotechnology and Bioengineering, 22,* 639–642 (1980).
25. Ohashi, M. T. W., T. Ishikawa, Y. Watanabe, K. Miwa, M. Shoda, Y. Ishikawa, T. Ando, T. Shibata, T. Kitsunai, N. Kamiyama, and Y. Oikawa, "Steam-Sterilizable Dissolved Oxygen Sensor and Cell Mass Sensor for On-Line Fermentation System Control," in *Computer Applications in Fermentation Technology,* W. B. Armiger, ed., Biotechnology and Bioengineering Symposium No. 9, pp. 103–116, John Wiley and Sons, New York, NY (1979).
26. Wang, D. I. C., I. Cooney, A. I. Demain, P. Donnill, A. E. Humphrey, and M. D. Lilly, "Fermentation Kinetics," in *Fermentation and Enzyme Technology,* John Wiley and Sons, Inc., New York (1979).
27. Erickson, R. and T. Holme, "The Use of Microcalorimetry as an Analytical Tool for Microbial Processes," in *Advances in Microbial Engineering,* B. S. Kyr, A. Prokoa, and M. Novak, eds., John Wiley & Sons, Inc., New York (1973).
28. Gregg, E. C. and K. D. Steidley, "Electrical Counting and Sizing of Mammalian Cells in Suspension," *J. Biophys., 5,* 393 (1960).
29. Gencer, M. A. and R. Mutharasan, "Determination of Biomass Concentration by Capacitance Measurement," *Biotechnology and Bioengineering, 21,* 1097–1103 (1979).
30. Kulys, J. and K. Kadzianskiene, "Yeast BOD Sensor," *Biotechnology and Bioengineering, 22,* 221–226 (1980).
31. Schultz, J. S. and G. Sims, "Affinity Sensors for Individual Metabolites," in *Computer Applications in Fermentation Technology,* W. B. Armiger, ed., Biotechnology and Bioengineering Symposium No. 9, pp. 65–71, John Wiley and Sons, Inc., New York, NY (1979).
32. Guilbault, G. G., "Ion-Selective Electrodes Applied to Enzyme Systems," *Ion Selective Electrode Rev., 4,* 187–231 (1982).
33. Guilbault, G. G., "Application of Enzyme Electrodes in Analysis," *Ann. New York Academy of Science, 369,* 285–294 (1981).
34. Mattiasson, B., B. Danielsson, G. F. Mandenius, and F. Winquist, "Enzyme Thermistors for Process Control," *Ann. New York Academy of Science, 369,* 295–305 (1981).

35. Hikuma, M., T. Yasuda, and I. Karube, "Application of Microbial Sensors to the Fermentation Process," *Ann. New York Academy of Science, 369,* 307–319 (1981).

36. Angell, J. B., S. C. Terry, and P. W. Barth, "Silicon Micromechanical Devices," *Sci. Amer., 284* (4), 44–55 (1983).

37. Chowdhury, J., "Shrinkage for a Bigger CPI Role," *Chemical Engineering, 21,* 18–23 (February 1983).

38. Howard, R. E., P. F. Liao, W. J. Skocpol, L. D. Jackel, and H. G. Craighead, "Microfabrication as a Scientific Tool," *Science, 221* (4606), 117–121 (1983).

39. Wolfe, R. and R. M. Dubrol, Jr., "Small Wonders: Microminiature Refrigerators for Cooling Detectors," *Photonic Spectra,* 52–54 (July 1983).

40. Vincent, A. and G. Priestley. *The Measurement of Process Variables, Topics in Enzyme and Fermentation Biotechnology,* Vol. 4, A. Wiseman, ed., John Wiley & Sons (1980).

41. Fleischaker, R. J., J. C. Weaver, and A. J. Sinskey, "Instrumentation for Process Control in Cell Culture," *Advances in Applied Microbiology, 27,* 137–167 (1981).

42. Tran, N. D., J. L. Romette, and D. Thomas, "An Enzyme Electrode for Specific Determination of L-Lysine: A Real Time Control Sensor," *Biotechnology and Bioengineering, 25,* 329–340 (1983).

43. Bar-Chaim, N., I. Ury, and A. Yariu, "Integrated Optoelectronics," *IEEE Spectrum,* 38–45 (May 1982).

44. Hilinski, E. F. and P. M. Rentzeris, "Biological Applications of Picosecond Spectroscopy,' *Nature, 307,* 481–487 (1983).

11

Chemical Field Effect Transistors (ChemFETs)

INTRODUCTION

B*iotechnology and microelectronics* share many properties. Both deal with the very small, both are high technologies, and both find many applications far from their original niches. Ion-selective field effect transistors (IsFETs) and chemical field effect transistors (ChemFETs) are the offspring of a late marriage between biochemistry and semiconductors, a union that is likely to result in significant progeny.

This chapter presents an overview of IsFETs and ChemFETs (referred to here collectively by the latter term). First, they are defined according to the three essential elements of which they are composed—semiconductors, enzymes, and membranes—and the interaction of those elements. Next, their biomedical, industrial, and defense applications are discussed. Finally, some new directions for this technology are suggested.

CHEMFETs

To understand how ChemFETs operate, we must examine their three building blocks—semiconductors, enzymes, and membranes.

Semiconductors

An essential element of ChemFETs is the metal oxide semiconductor field effect transistor (MOSFET), an example of which is diagrammed in Figure 11-1.

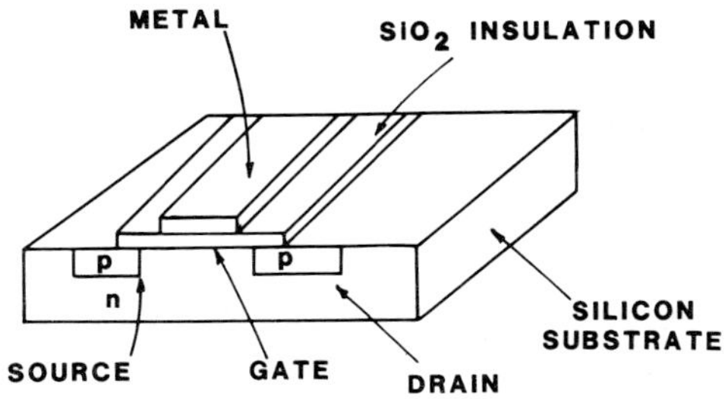

Figure 11-1. A cross-section of a p-channel MOSFET.

Insulators do not conduct electricity. However, if a large amount of energy is added, an insulator can be made to conduct. The electrons in an insulator are held, energy wide, in the valence band; if electrons are to jump the forbidden gap into the conduction band, a large amount of energy must be applied. Semiconductors fall into a category between insulators (large forbidden gap) and conductors (narrow forbidden gap). In a semiconductor, the silicon crystal is held together by covalent bonds that arise from the sharing of electrons between neighbors; this forms a configuration that is energically favorable.

To give a semiconductor interesting conductivity, it is doped with impurities in the form of boron and phosphorus. When a phosphorus atom replaces a silicon atom, an extra electron becomes available. The semiconductor is then said to be of the n-type (presence of a negative carrier, or electron). The doping favors electron conduction in the conduction band. In the presence of boron, we have the reverse case: the carriers are positively charged, or holes, and we have a p-type semiconductor. Acceptor impurities creating positive sites will generate a new energy level slightly above the valence band; electrons from donor impurities are located, energy wise, below the conduction band.

When a p-n junction is established, the electrons from the n-crystals are attracted across the junction to fill holes in the p-crystal. In turn, some holes from the p-crystal are attracted across the junction to the electrons in the n-crystal. That is, diffusion takes place, causing an area on both sides of the junction to become void of current carriers. This space is called the depletion area. The diffusion of electrons and holes results in the development of an electrical potential—the so-called potential barrier. Note that the overall charge of the p-n diode has not changed.

If a negative voltage is applied to the p-crystal and a positive charge to the n-crystal, the depletion area will widen through the increased migration of electrons to the positive source and of holes to the negative source. The potential barrier and the source voltage work together to prevent the current carriers from crossing the junction. The diode is said to be reverse biased, and there is only a small leakage current.

If the diode is connected in a forward bias mode (positive voltage to the p-crystal and negative voltage to the n-crystal), the depletion area decreases in width, the potential barrier is considerably reduced, and current flows through the junction. This is because, in this configuration, electrons from the p-crystals are attracted to the positive source. For each electron leaving the p-crystal, a hole is created in the p-material. These holes drift toward the junction, where they are filled with electrons from the n-crystal.

The field effect transistor in which an n-material is diffused in a p-substrate is then a current amplifier. The FET depends on majority carriers (electrons) for conduction between source and drain. A small change in voltage at the gate will cause a large drain current.

If an n-channel of moderate resistivity is diffused between the source and the drain, current can flow when the gate potential is zero. For a positive gate voltage, the device enhances carriers in the channel; for a negative gate voltage, the channel is depleted of carriers. MOSFETs can be operated in both depletion and enhancement modes. The depletion mode refers to the decrease of carriers in the channel due to variation in the gate voltage; the depletion mode device has considerable drain current flow for zero gate voltage. The enhancement mode refers to the increase of carriers in the channel due to gate voltage (See Figure 11-2).

The sensitivity of a semiconductor is measured practically by a figure of merit called the transconductance or forward transadmittance. This is a relation between input voltage and output current for a constant source-drain voltage [1,2]:

$$g_m = \Delta I_D / \Delta V_G$$

Where I_D is the drain current and V_G is the gate voltage. At saturation, where $|V_D| = |V_G - V_T|$, we have

$$I_D = K\,(V_G - V_T)^2$$

where V_T is the threshold voltage, or the voltage necessary to terminate conduction (in the depletion mode) or initiate conduction (in the enhancement mode). Thus

$$g_m = 2K\,(V_G - V_T)$$

$$\text{and } K = \frac{\mu \varepsilon_{ox}\, \varepsilon_o W}{2\, t_{ox}\, L}$$

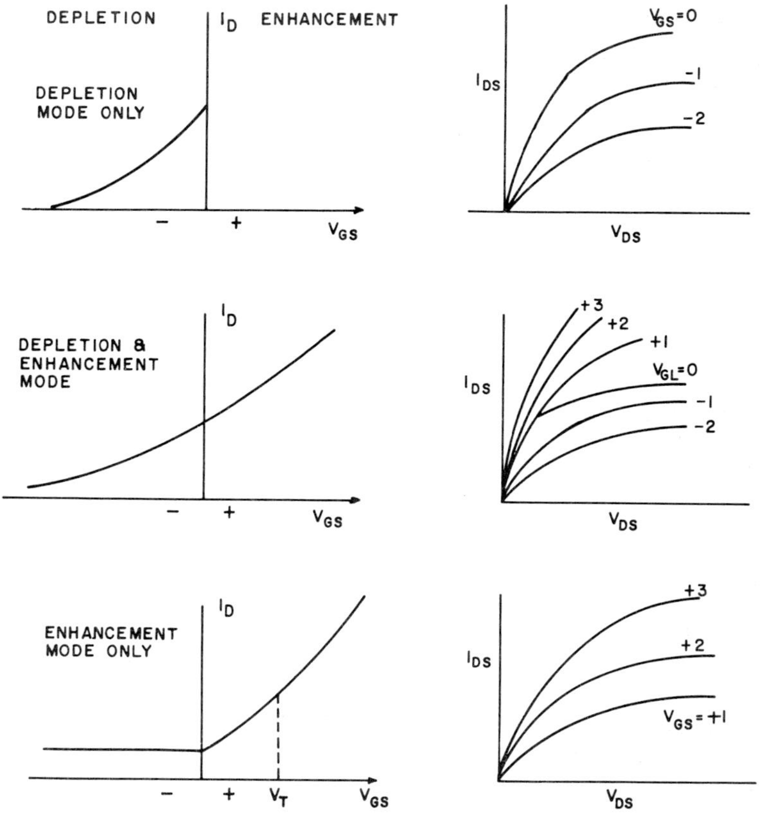

Figure 11-2. Characteristics of the various MOSFET modes of operation.

Where W is the width of the gate region, L is its length, μ is the electron mobility in the channel, ε_o is the dielectric constant, and t_{ox} is the gate oxide thickness.

The threshold voltage is given by

$$V_T = \phi_{ms} - \frac{Q_{ss}}{C_o} + 2\phi_F - \frac{Q_B}{C_o}$$

voltage required to establish a flat band condition	voltage required to bend the band

where ϕ_{ms} is the work function of the gate electrode; ϕ_F is the Fermi potential; Q_{ss} is the charge at the silicon-silicon oxide interface; Q_B is the charge induced in the surface depletion layer generated by the ions on the surface; and C_o is the gate capacitance: $C_o = \varepsilon_{ox}\, \varepsilon_o / t_{ox}$.

The change in threshold voltage is

$$\Delta V_T = \Delta \Omega_{ss} / C_o$$

$$\Delta I_D = g_m\, \Delta V_T$$

and

$$\Delta I_D = g_m\, \Delta Q_{ss} / C_o$$

Using more or less typical values of the design parameters for a p-channel in the enhancement mode and applying a given voltage would give a value of g_m as shown below:

$W = 0.37 \times 10^{-2}$ cm	$L = 0.015 \times 10^{-2}$ cm
$\varepsilon_{ox} = 4$	$\varepsilon_{ox} = 8.85 \times 10^{-14}$ F/cm
$t_{ox} = 1000$Å	$\mu = 190$ cm^2/V sec
$V_G = -10$ V	$V_T = -4$ V
$V_D = -1$ V	

Elementary arithmetic yields $g_m = 34 \times 10^{-6}$ mhos.

Playing with these parameters permits the design of MOSFETs to satisfy detection needs.

Enzymes

Enzymes are organic molecules, biocatalysts of extreme specificity and activity. They have a number of useful characteristics that permit them to do the following:

- Operate effectively at mild temperatures
- Perform in aqueous solutions
- Have a very high specificity
- Require a low concentration of substrate, can detect and react with small quantities of substrate
- Speed up chemical reaction by many orders of magnitude
- Be readily available in nature (plants, animals, bacteria, yeast, fungi)
- Catalyze virtually every reaction known to man:
  —oxidation
  —acylation
  —reduction
  —phosphorylation
  —racemisation
  —sulfurization

—decarboxilation
—desamination
—hydrolysis
—side chain addition

The use of enzymes is not without difficulty: purification is time-consuming and costly; many enzymes require expensive cofactors; and in most cases, enzymes are rapidly inactivated outside a narrow range of pH and temperature. In spite of their attractive characteristics, enzymes are a relatively small market in the U.S., some $150 million. The biggest share of this market is taken by amylases (e.g., glucoamylase), proteases (e.g., rennin), glucose isomerase, invertase, cellulase, and pectinases.

Enzymes can perform a variety of functions as indicated by the following examples:

Catalase—catalyzes the decomposition of H_2O_2 to water and oxygen
α amylase—hydrolyzes α-(1–4) glucan linkages of starch
Pepsin—hydrolyzes peptides
Glucose oxidase—oxidizes glucose to gluconic acid in the presence of oxygen
Glucose isomerase—converts aldo sugars to their keto isomers
Invertase—hydrolyzes β-D-fructofuranosyl linkages and converts sucrose to glucose and fructose

In the medical world, enzymes used as detectors and targets of testing reactions represent some 25% of today's clinical laboratory workload. Because enzyme modifications and changes are associated with many disease states, they are very good indicators of our health and of the harmonious functioning of our organs, tissues, and cells. Enzymes are also used in therapy, although they represent less than 1% of all prescribed drugs and medications. The use of enzymes in therapy is a new field. Hyaluronidase, which hydrolyzes hyaluronic acid, a polysaccharide found in tissues, can be used to increase the absorption and dispersion of injected drugs. Enzyme deficiency due to genetic defect can be corrected by the administration and delivery of the missing enzyme. Urokinase has been shown to act on the human fibrinolytic mechanism, accelerating the dissolution of pulmonary emboli. The restriction of asparagine by administration of L-asparaginase has been used as a selective treatment for leukemia.

In contrast, there are very few examples of enzymes used as monitors or indicators of process conditions in the industrial environment.

Membranes

The presence of membranes to compartmentalize cells, structures, or reactions is a characteristic of life itself. The fundamental unit of cell function is the lipid bilayer membrane. Impermeable to water, it is a good insulator, and many of its constituent proteins provide pores or pathways for passive and active transport of ions and molecules in both directions. The membrane is an effective electron transport system. Biological membranes, although diverse in shape and function, are surprisingly similar. They are made of lipids, proteins, and carbohydrates, with the amphipathic lipid molecules responsible for membrane integrity. These molecules are organized in two hydrophilic surfaces separated by a hydrophobic core.

It is not surprising that, in designing a wide variety of membranes for industrial applications, engineers have been for years copying and sometimes improving on nature. Based on a technique developed by Irving Langmuir and Kathleen Blodgett, it is possible to engineer thin films of organic material in molecular layers. It is today theoretically possible to engineer a membrane to isolate, separate, concentrate, or purify any single chemical species in a complex, hostile environment.

The membranes presently in use can be classified into two broad categories, based on the material which they are made—organic and inorganic. Organic membranes can be further separated into ionic and non-ionic. Ionic membranes can be selective about the material that passes through them by discriminating against the electrical charge (positive or negative) associated with that material; non-ionic membranes have stronger structures and discriminate against size only. Organic membranes have a distinct disadvantage in that they cannot operate at extreme temperature and pH conditions, unless one of these variables is made less severe. Inorganic membranes do not have this limitation.

Synthetic membranes can be further classified by their ability to discriminate between substances on the basis of several selection properties: molecular size, electric charge, and different affinity for two dissimilar solvents.

Molecular Size

Tortuous-pore membranes consist of a polymer matrix surrounding numerous interconnected vacuoles, resulting in convoluted pathways through the membrane matrix material. The most common technique for manufacture of tortuous-pore microfiltration membranes involves a phase-inversion casting process in which the relative amounts of polymer, solvent, and water, as well as the drying rates, are all well controlled to determine the number and size of the vacuoles in the polymer matrix. In contrast, capillary-pore membranes are the result of a novel technique that produces straight-channel cylindrical pores with uniform diameter.

Electric Charge

Charged membranes can easily discriminate between ions according to their electric charge. A negatively charged membrane made of a sulfonated polymer with the sulfonate group (SO_3^-) strategically placed along the walls of the core will allow positive (e.g., Na^+) ions to pass, blocking negative ions (e.g., Cl^-). A positively charged membrane based on a derivative of the quarternary-ammonium (NH_4^+) operates the opposite way.

Solvent Affinity

The bipolar membrane is double, having a cation-exchange membrane on one side and an anion-exchange membrane on the other side. These two membranes are glued together, with a thin layer of solution (e.g., water) left between. This solution disassociates, and the bipolar membrane acts as an electrode. On the other hand, facilitated transport membranes are solvent extraction membranes. A specialized transport molecule is incorporated within the membrane, forming a complex with the species of interest and transporting it across the membrane. Some membranes are made of a microporous support that contains a liquid confined by capillary action. The infinite selectivity of such membranes is achieved by adding specific carriers, typically ion-exchange, which shuttle metals from one side of the membrane to the other. Such membranes have been used in ion-specific electrodes.

Still another way to distinguish membranes is by their form. Membranes can be packaged in the form of a sheet, wound up in a tight tube, or spun in the form of a hollow fiber.

Liquid membranes can be used to sequester chemicals and are virtually microscopic reaction vessels of great flexibility. Liquids are known to exhibit extreme selectivities between chemically related species. Liquid membranes fall into two categories: support-type and emulsion-type. The former utilizes a microporous membrane whose pores are filled with liquid. The latter, also known as a liquid-surfactant membrane (emulsion type), can be operational when the membrane forms tiny droplets suspended in a liquid phase. Liquid membranes are capable of moving a particular solute from a region of high concentration into a region of low concentration. A liquid membrane can separate two miscible liquids and can control the mass transfer between them.

Three main types of diffusion can occur through liquid membranes:

- Simple passive diffusion down a gradient in concentration or thermodynamic activity.
- Facilitated transport, where a carrier agent capable of combining with the permeate to be transported is dissolved in the liquid membrane phase.
- Coupled transport, in which the carrier agent couples the flow of two or more species; because of this coupling, one of the species can be moved against its concentration gradient, as long as the concentration gradient of the second coupled species is sufficiently large.

In addition, a "facilitated transport or carrier transport" mechanism can be used to enhance the mass transfer. This is achieved by incorporating one of the following: a reagent in the encapsulated phase that can react with the permeating compound from the continuous aqueous phase; or a compound, such as a complexing agent, in the membrane phase to increase the solubility of the permeating compound in the membrane, thus increasing the permeation rate of this compound through the membrane.

The design of a liquid membrane process requires technical expertise, and can be tailored to specific applications. The separation mechanics are similar to a two-stage, liquid-liquid extraction. However, the separation becomes more complicated when multiple-stage extraction is attempted. Liquid membranes have some of the characteristics of biological membranes. Like biomembranes, they can be made highly fluid and, because of their low viscosity, they exhibit low diffusivities; both types of membranes can behave as "chemical pumps."

In some cases, a membrane serves as support for enzymes, cells, or bacteria. Cells or bacteria can be sequestered behind or sandwiched between membranes, and the nutrient flow used to control, stimulate, or inhibit the growth of the organisms. The immobilization of enzymes unto and into membranes permits us to take advantage of their great activity, selectivity, specificity, and sensitivity.

Two final points about membranes might be noted. First, this discussion may have given the impression that membranes are used only to separate solids from liquids; in fact, they have been used in gas separation and enrichment. Second, the case for membranes should not be overstated. Much secrecy surrounds their fabrication, and a great deal of art must be applied; they are sensitive to fouling and can be very fragile.

Un Mariage a Trois

Combining the three building blocks described above—the semiconductor, the enzyme, and the membrane—is an engineering feat. ChemFETs are a refinement of an older technology—enzyme electrodes [3–8]. More than 60 enzyme electrode probes have been described in the literature, many of which are available commercially. Many are totally specific for a single chemical. The substrates measured include sugars (glucose, sucrose, maltose, lactose); acids (acetic, uric, formic, oxalic, lactic); amino acids (arginine, asparagine, cysteine, glutamine, glutamic acid, histidine, lysine, phenylalanine, tyrosine); urea; cholesterol; creatinine; penicilin;and acetylcholine.

Microbial electrode sensors have been developed. With this technology, the whole cell is immobilized, and a product of the reaction is measured. A few examples include the following:

Substrate	Immobilized Microorganisms	Sensor
BOD	*Trichosporon cutaneum*	O_2 probe
Ammonia	Nitrifying bacteria	O_2 probe
Acetic acid	*Trichosporon brassiere*	O_2 probe
Ethyl alcohol	*Trichosporon brassiere*	O_2 probe
Glutamic acid	*Escherichia coli*	CO_2 probe

A related concept is the enzyme thermistor [9–11]. When a substrate is fed to an enzyme, a product is formed with the liberation of heat. A thermistor is used to monitor this heat of reaction, which can be correlated with the concentration of the substrate.

A body of knowledge is slowly accumulating on the properties, behavior, and fabrication theory and practices of chemical field effect transistors [12–26]. More extensive knowledge of surface chemistry phenomena and electron transport activities is required before success can be achieved [27–28].

The schematics for a MOSFET and a ChemFET are shown side by side (Figure 11-3) to emphasize their similarities and the differences. Note that in a ChemFET, the gate has been replaced by a membrane containing an enzyme; the solution to be analyzed sits on top of the membrane, and a reference electrode completes the assembly.

Adsorption in the gate film or diffusion through it of polar molecules or ions results in the establishment of a localized field; this causes the semiconductor to accumulate charges to balance or offset the field. These charges modulate the flow of carriers in the channel, near the surface, between the two n-regions. Several groups have developed pH-sensitive field effect transistors that are accurate, sensitive, and effi-

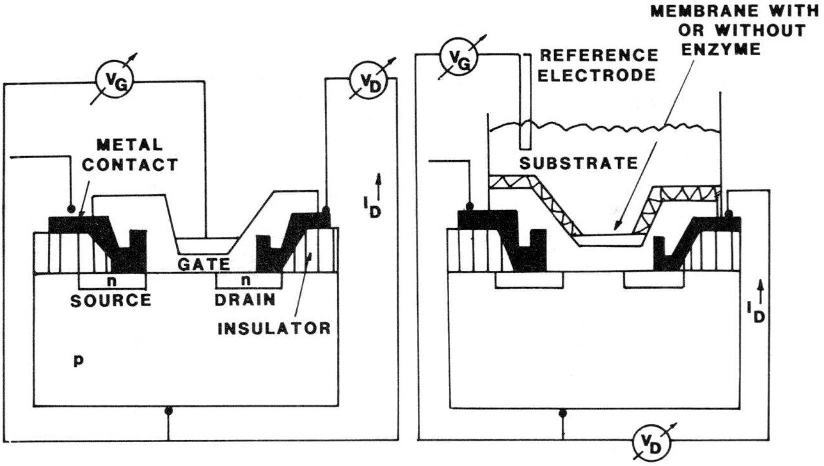

Figure 11-3. Comparison of MOSFET and ChemFET.

cient. For these, bare-gate insulators (e.g., silicon nitride) are used [29–32].

To take advantage of the great sensitivity of semiconductors to the detection of ionic species, it became necessary to ensure that the devices were extremely selective, further, the reaction of interest had to be reversible so that the device could be reused many times. These characteristics are achieved through the use of removable selective membranes. Polymeric membranes selective to specific ions (e.g., valinomycin diocytalipate for potassium; nonactin/monactin, tris (2-ethylhexyl) phosphate, PVC for NH_4^+; ionophore ETH for Na^+) can be applied to the insulating layer of the FET [33,34]. Heterogeneous membranes (inorganic salt in an elastomer) have been applied to detect additional species (NO_3^-, SO_4^-, Cl^-, Br^-, I^-) [33–35].

The classic example of the use of a ChemFET is the detection of glucose by the action of the enzyme glucose oxidase. A change in pH (due to gluconic acid) or the consumption or production of oxygen (by the use of a second enzyme-catalase) can all be used as end points:

$$\text{Glucose} + O_2 \xrightarrow{\text{glucose oxidase}} \text{gluconic acid} + H_2O_2$$

$$H_2O_2 \xrightarrow{\text{catalase}} \tfrac{1}{2}\, O_2 + H_2O$$

A number of enzymes are known to produce or consume H_2, NH_3, O_2, or CO_2 during their reaction, and are good candidates for detection.

In another application, instead of the metallic gate electrode of conventional FETs, a monomolecular layer of an antibody protein specific to an antigenic protein to be detected can be adsorbed onto the insulating layer. The charge at this gate region is altered by the reaction (electrostatically mediated) between the antigen and antibody, thus alerting the drain to source current in the transistor. The time rate of change of this current is a measure of the concentration in solution of the antigenic protein being detected. Viruses, bacteria, and a variety of cells with appropriate markers can be detected this way. This arrangement has encountered difficulties associated with immobilizing the antibody while retaining activity, and with interfering ions [36–39]. These problems might be resolved by the use of monoclonal antibody and a method similar to enzyme-linked immunoassay (where enzyme activity detection is the surrogate measure for the antibody-antigen reaction) [40–45].

The membrane can be a passive support for the reacting enzyme; in a more sophisticated arrangement, it can perform a selection function for a substrate to be measured or the product of a reaction.

Numerous combinations of substrates, enzymes, membranes, reaction milieus, products, and detectors are possible. Figure 11-4 shows just one example. The semiconductor can be coated with a great variety of chemical agents that become activated by another species during the reaction. The membrane can serve as support, as a selection and transport vehicle, or as a reaction vessel. The chain from substrate to detector can be expanded by having additional reaction vessels sandwiched in-between. Only imagination and creativity limit what can be achieved.

In practice, substantial difficulties have been encoutered in using ChemFETs (for other purpose than pH or small ion measurements). The theoretical knowledge on surface chemistry and electrochemical phenomena at interface is very thin, and difficulties will continue to be encountered until a more substantial knowledge base has been achieved.

APPLICATIONS

The range of ChemFET applications is wide—from medical diagnosis to pollution monitoring. A number of ions (Ca^{2+}, H^+, NO_3^-, SO_4^-, Cl^-, K^+, Na^+, Br^-, I^-) and chemical species (glucose, urea, creatine, penicillin, organophosphorus) have been successfully measured using such devices [46].

Many organizations are involved in research and development. These include ITT, the University of Utah, Johnson & Johnson, the Universi-

University of Pennsylvania. Some potential biomedical, industrial, environmental, and defense applications are discussed below.

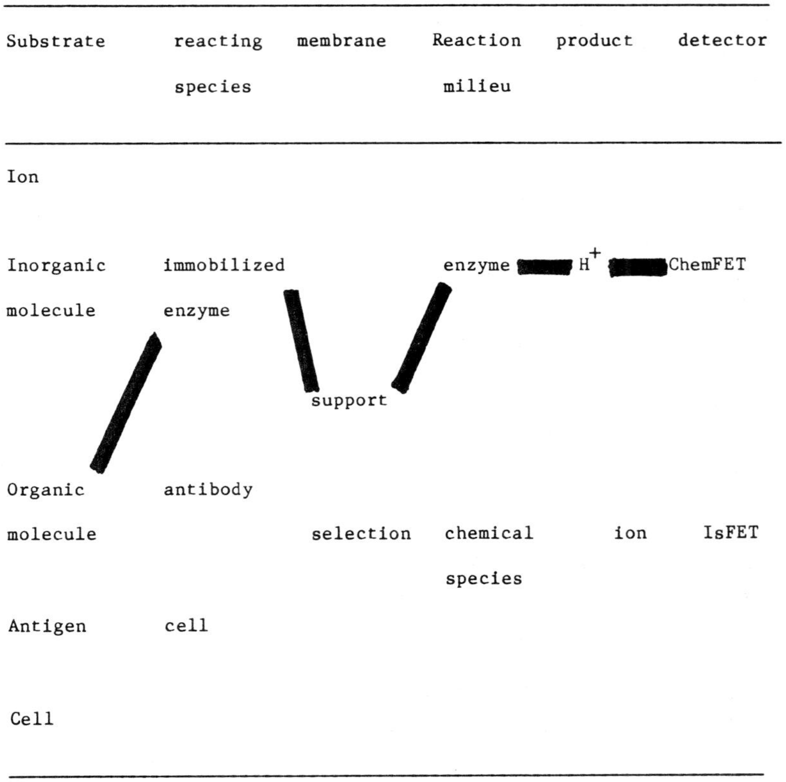

Figure 11-4. A typical path for the detection of glucose by oxidase immobilized on a membrane and producing gluconic acid detected by a ChemFET is shown.

ty of Warwick, the University of Wales, the Bendix Company, the University of Michigan, Case Western University, and the University of Pennsylvania. Some potential biomedical, industrial, environmental, and defense applications are discussed below.

Biomedical Applications

The human body contains a variety of ions, small molecules, proteins, hormones, etc. With creativity and the proper combination of semiconductors, enzymes and membranes, a wide variety of species can be detected to assist in diagnosis or treatment. The following are some examples:

- Detection of antibodies and antigens associated with infectious disease. (To measure antibody concentration, the gate region of the device would be coated with antigen rather than enzyme.)
- Measurement of hormone concentrations to detect endocrine system disorders. Several substances could be measured in healthy patients and under a variety of disease conditions. These include hormones such as insulin, glucagon (sugar metabolism); renin-angiotensin (blood pressure); parathyroid hormone and calcitonin (calcium regulation); chorionic gonadotrophin, luteinising; and follicle-stimulating hormones (reproduction); prolactin; growth hormones (growth regulation); thyroid-stimulating hormone; gastrointestinal hormones; and pituitary adrenocorticothophin (cortisol, aldosterone secretion).
- Mass screening of patients to detect patterns in serum protein levels, or electrolytes.

The possible development of a multisensor probe for intra-orifices and intravenous positioning would be invaluable to clinical medicine. An intra-oral probe to detect pH and ions is now under development. A Chemfet attached to the end of a catheter was used intravenously in animals to measure potassium concentration [46].

Industrial Applications

The greatest opportunities for ChemFET's are in the industrial world, where a knowledge of feedstock quality, process conditions, and product quality is required in real time to effect process control.

As an example, control of the fermentation processes is dependent on frequent monitoring of a number of factors: pH, temperature, dissolved oxygen, redox potential, substrate concentrations, mass flow rate, heat evolution, gaseous oxygen and carbon dioxide concentrations, cell mass, enzyme levels, cofactor activity, and the amount of product in the reactor under very heterogeneous conditions. In the production of penicillin, enzyme electrodes have been developed in which the enzyme penicillinase is immobilized over a pH glass electrode. Penicillin is measured indirectly by detecting the reduction in pH associated with the penicillinase-catalyzed hydrolysis of penicillin to penicilloic acid. More recently, this concept has been adapted to FET technology by depositing penicillinase over a pH-sensitive FET. The device consists of two separate gate regions—a penicillin-sensitive gate (coated with the enzyme) and a reference gate that exhibits only pH response. Enzyme-based FETs similar to those described above could be developed to detect other, selected fermentation products [47].

It is possible to predict the existence, in a few years, of a whole family of enzyme electrodes, IsFETs, ChemFETs, pyroelectric devices, and the like that would simultaneously sense key parameters in real time and feed this information to minicomputers. This would permit modeling of the process, optimization of operating conditions, and control of the operations.

Another proposed industrial application is the detection of corrosion in pipes and oil storage tanks. Corrosion is often due to bacteria that reduce sulfates to sulfide ions. In this application, bacteria would be imbedded in the membrane surrounding the sensing fluid, permitting the FET to detect the sulfide ion potentials released during this reaction. Inserted in the pipe or tank, this device could provide early detection of corrosion without the cost of physical inspection.

Environmental Applications

Environmental protection requires monitoring, often involving the detection of small quantities of substances in a variety of media (air, water, soil) under various conditions of temperature, humidity, pressure, and interfering substances. Advanced analytical systems, such as combined gas chromatography/mass spectrometry, have improved the sensitivity with which these measurements can be made, and increased the number of substances which can be readily detected; however, these technologies remain costly and difficult to apply under field conditions. The advantages of FETs—their sensitivity, size, ruggedness, and simplicity of operation—would seem applicable to environmental sensing. For example, a probe with multiple ion or molecular detection capability, combined with an integrated circuit chip, could be designed to detect and interpret chemical "signatures" of interest (i.e., the groups of compounds identified with particular pollution sources). Parameters of interest go beyond the pH and small ions measured up to now, and would include the following [48]:

- Detection of trace amounts of toxic substances in air and water (e.g., volatilized toxic substances in the air over a hazardous waste site), or in the adjacent water streams
- Industrial hygiene monitoring (e.g., a potentially more sensitive replacement for the badges worn by workers to detect exposure to hazardous substances)
- Indoor air pollution—(e.g., the manufacture of low-cost sensors for detecting trace air pollutants in home and office environments)

The combination of low cost, small size, ruggedness, and capability for multiple detection and signal processing should lead to increased R&D into the uses of ChemFETs in environmental monitoring.

Defense Applications

The U.S. Department of Defense (DoD) has been one of the major sponsors of research in FET sensor development. DoD has sponsored studies on the use of enzyme systems for alarms indicating the presence of organophosphate nerve agents. The toxic effect of these agents is based on the inhibition of the cholinesterase enzyme. Early work was directed toward the development of immobilized enzyme electrodes that could meet the following criteria [45]:

- Sensitivity of 1 part per billion (ppb)
- Response time of less than 20 seconds
- High selectivity of response (i.e., no false alarms)

Although enzyme electrodes meeting these criteria have been developed, recent attention has turned to the application of ChemFETs, including an enzyme-coupled device based on the same principle of cholinesterase inhibition (the transistor detects the alerted potential caused by changing enzyme activity, which, in turn, is a result of inhibition by the organophosphate compounds). It was found that the advantages of small size, high sensitivity, and low cost must be balanced against the disadvantages associated with the irreversibility of the enzyme reaction and the limited usable lifetime (hours rather than days, although future improvements in shelf life are likely). A synthetic enzyme is being put to work to perform the cholinesterase reaction.

As with civilian applications, the possibilities for defense applications extend to any chemical monitoring situation in which sensitivity, low cost, small size, and reliability are useful. In addition to the detection of chemical and biological agents under battlefield conditions, biosensors might be used in the future verification monitoring of agreements to limit the development and use of such weapons.

NEW DIRECTIONS

A good thing can always be improved. While metal oxide semiconductors will be enhanced, new efforts are being directed toward electronic conducting polymers, especially the organic type. These metal-like organic polymers provide a variety of semiconductors. Semiconductor materials have been synthesized using polyacetylene, poly (p-phenylene), and poly (p-phenylene sulfide) via polymer doping or matrix-controlled solid-state polymerization.

New developments in membranes are merely a question of imagination, creativity, and more difficult engineering.

New enzymes with interesting properties are added to our reservoir of knowledge almost daily. If we cannot find a new one, we shall invent it. A number of teams are hard at work building synthetic enzymes capable of speeding reactions a billion fold.

The secret of any successful industrial process is knowledge of the process conditions (through monitoring) and process control. There are promising applications involving the detection of changing variables through ChemFETs, with signals fed to an existing process control computer.

ChemFETs are promising potentiometric detection devices, although progress in related technologies will occur only if we improve our knowledge of the interface between solid-state electronics and living systems. Enzyme electrodes are a competing well-established technology. Methods other than potentiometry, such as nanocalorimetry (pyroelectric), are likely to develop in parallel. Eventually, a whole arsenal of bimolecules-to-electronic output devices will appear on the market, replacing a large segment of today's detectors.

REFERENCES

1. Kane, J. F. and D. L. Wollesen, "Field-Effect Transistors in Theory and Practice," *Motorola Monitor, AN-211,* 1–12.
2. Schenck, J. F., "A Field Effect Transistor Modified for Use as a Sensor of Ions in Solution," General Electric, Report No. 77CRD063 (1977).
3. Guilbault, G. G., "Applications of Enzyme Electrodes in Analysis," in *Biochemical Engineering II,* A. Constantinides, W. R. Vieth, and K. Venkatasubramanian, eds., The New York Academy of Sciences, New York (1981).
4. Guilbault, G. G., "Analytical Uses of Immobilized Enzymes," in *Enzyme Engineering,* Vol. 6, I. Chibata, S. Fukui, and L. B. Wingard, Jr., eds., Plenum Press, New York (1982).
5. Guilbault, G. G., "Analytical Uses of Immobilized Enzymes," in *Enzyme Engineering,* Vol. 2, E. K. Pye and L. B. Wingard, Jr., eds., Plenum Press, New York (1974).
6. Guilbault, G. G., "Applications of Enzyme Electrodes in Analysis," *Annals New York Academy of Sciences, 369,* 285–294 (1981).
7. Chern, G.-C. and J. N. Zemel, "Temperature Dependence of the Gate-Controlled Portion of Ion-Controlled Diodes," *IEEE Transactions on Electron Devices, 29* (1), 115–123 (1982).

8. Rawls, R. L., "Electrodes Hold Promise in Biomedical Uses," *C&EN,* 19–20 (Jan. 5, 1976).

9. Danielsson, B., I. Lundstrom, K. Mosbach, and L. Stiblert, "On a New Enzyme Transducer Combination: The Enzyme Transistor," *Analytical Letters, 12* (B11), 1189–1199 (1979).

10. Mattiasson, B., B. Danielsson, C. F. Mandenius, and F. Winquist, "Enzyme Thermistors for Process Control," in *Biochemical Engineering II,* A. Constantinides, W. R. Vieth, and K. Venkatasubramanian, eds., The New York Academy of Sciences, New York, NY (1981).

11. Cooney, C. L., J. C. Weaver, S. R. Tannenbaum, D. V. Faller, A. Shields, and M. Jahnke, "The Thermal Enzyme Probe—A Novel Approach to Chemical Analysis," *Enzyme Engineering,* Vol. 2, E. K. Pye and L. B. Wingard, Jr., eds., Plenum Press, New York (1974).

12. Zemel, J. N., "Chemically Sensitive Semiconductor Devices," *Research/Development,* 38–44 (April 1977).

13. Haemmerli, A., J. Janata, and J. J. Brophy, "Equilibrium Noise in Ion Selective Field Effect Transistors," *Jnl. of the Electrochemical Society, 129* (10), 2306–2312 (1982).

14. Zemel, J. N., "Ion-Sensitive Field Effect Transistors and Related Devices," *Analytical Chemistry, 47* (2), 255A–266A (1975).

15. Lauks, I., "Polarizable Electrodes," *Sensors and Actuators, 1,* 261–268 (1981).

16. Lauks, I., "Polarizable Electrodes, Part II," *Sensors and Actuators, 1,* 393–402 (1981).

17. Keramati, B. and J. N. Zemel, "Pd-Thin-SiO_2-Si Diode. I. Isothermal Variation of H_2-Induced Interfacial Trapping States," *J. Appl. Phys., 53* (2), 1091–1099 (1982).

18. Keramati, B. and J. N. Zemel, "Pd-Thin-SiO_2-Si Diode. II. Theoretical Modeling and the H_2 Response," *J. Appl. Phys., 53* (2), 1100–1109 (1982).

19. Akers, L. and T. M. Wang, "Simulation of a Three Dimensional Semiconductor Device," *Simulation,* 43–50 (Feb. 1983).

20. Vosburg, K. G. and J. F. Schenck, "Solid State Surface Charge Sensor,' *Rev. Sci. Instrument, 47* (8), 972 (1976).

21. Bergveld, P. and N. F. DeRoois, "Physical Mechanisms for Chemically Sensitive Semiconductor Devices," *Nature, 273,* 438–443 (1978).

22. Chern, G.-C. and J. N. Zemel, "Temperature Dependence of the Gate-Controlled Portion of Ion-Controlled Diodes," *IEEE Transactions on Electron Devices, Ed29* (1), 115–123 (1982).

23. Buck, R. P., "Kinetics and Drift of Gate Voltages for Electrolyte-Bathed Chemically Sensitive Semiconductor Devices," *IEEE Transactions on Electron Devices, Ed29* (1), 108–113 (1982).

24. Schenck, J. F., "Application of Electronic and Colloid Science Concepts to the Formulation IsFET Device Physics," in *Theory, Design and Biomedical Applications of Solid State Chemical Sensors,* CRC Press, West Palm Beach (1978).

25. Buck, R. D. and D. E. Hackleman, "Field Effect Potentiometric Sensors," *Analytical Chemistry, 49* (14), 2315–2321 (1977).

26. Danielson, B., I. Lundstrom, K. Mosbach, and L. Stilbert, "A New Enzyme

Transducer Combination, The Enzyme Transistor," *Anal. Letter, 12* (311), 1185–1189 (1979).

27. Schenck, J. F., "Technical Difficulties Remaining to the Application of IsFET Devices," in *Theory, Design and Biomedical Applications of Solid State Chemical Sensors,* CRC Press, West Palm Beach (1978).

28. Schenck, J. F., "A Transistor Method for Measuring Changes in Double Layer Potentials," *J. Colloid and Interface Sci., 61* (3), 569–576 (1975).

29. Lauks, I. R., "pH Measurements Using Polarizable Electrodes," *IEEE Transactions on Electron Devices, 26* (12), 1952–1959 (1979).

30. Akiyama, T., Y. Ujihira, Y. Okabe, T. Sugano, and E. Niki, "Ion-Sensitive Field-Effect Transistors with Inorganic Gate Oxide for pH Sensing," *IEEE Transactions on Electron Devices, Ed29* (12), 1936–1941 (1982).

31. Katsube, T., I. Lauks, and J. N. Zemel, "pH-Sensitive Sputtered Iridium Oxide Films," *Sensors and Actuators, 2,* 399–410 (1982).

32. Oesch, U., S. Caras, and J. Janata, "Field Effect Transistors Sensitive to Sodium and Ammonium Ions," *Anal. Chem., 53,* 1983–1986 (1981).

33. McKinley, B. A., J. Saffle, W. S. Jordan, J. Janata, S. D. Moss, and D. R. Westenskow, "In vivo Continuous Monitoring of K^+ in Animals Using ISFET Probes," *Medical Instrumentation, 14* (2), 93–97 (1980).

34. Janata, J. and R. J. Huber, "Ion-Sensitive Field Effect Transistors," *Ion-Sensitive Electrode Rev., 1,* 31–79 (1981).

35. Lauks, I. R. and J. N. Zemel, "The Si_3N_4/Si Ion-Sensitive Semiconductor Electrode," *IEEE Transactions on Electron Devices, 26* (12), 1959–1964 (1979).

36. Janata, J., "An Immunoelectrode," *Jnl. Amer. Chem. Soc., 97* (10), 2914–2916 (1975).

37. Giaever, I., "Method and Apparatus for Immunological Detection of Biological Particles," U.S. Patent No. 3,853,467 (Dec. 10, 1974).

38. Schenck, J. F., "Field Effect Transistor for Detection of Biological Reactions," U.S. Patent No. 4,238,757 (Dec. 9, 1980).

39. Mattiasson, B. and H. Nilsson, "An Enzyme Immunoelectrode," *FEBS Letters, 78* (2), 251–254 (1977).

40. Gebauer, C. R. and G. A. Rechnitz, "Immunoassay Studies Using Adenosine Deaminase Enzyme with Potentiometric Rate Measurement," *Analytical Letters, 14* (B2), 97–109 (1981).

41. Aizawa, M., S. Kato, and S. Suzuki, "Immunoresponsive Membrane I. Membrane Potential Change Associated with an Immunochemical Reaction Between Membrane-Bound Antigen and Free Antibody," *Jnl. of Membrane Science, 2,* 125–132 (1977).

42. Meyerhoff, M. E. and G. A. Rechnitz, "Electrode-Based Enzyme Immunoassays Using Urea Conjugates," *Analytical Biochemistry, 95,* 483–493 (1979).

43. Biotieux, J.-L., G. Desmet, and D. Thomas, "An 'Antibody Electrode,' Preliminary Report on a New Approach in Enzyme Immunoassay," *Clinical Chemistry, 25* (2), 318–321 (1979).

44. Blackburn, G. and J. Janata, "The Suspended Mesh Ion Selective Field Effect Transistor," *J. Electrochem. Soc., 129* (11), 2580–2584 (1982).

45. Janata, J., R. J. Huber, R. Cohen, and E. S. Kolesar, Jr., "Chemically Sensitive

Field-Effect Transistor to Detect Organophosphorous Compounds and Pesticides," *Aviation, Space, and Environmental Medicine, 52* (11), 666–671 (1981).

46. Yanchinski, S., "Biochips Speed Up Chemical Analysis," *New Scientist,* 236 (Jan. 28, 1982).
47. Caras, S. and J. Janata, "Field Effect Transistor Sensitive to Penicillin," *Anal. Chem., 52,* 1935–1937 (1980).
48. Kember, P. N., "IsFETS—New Tools for Water Quality Monitoring," *Wat. Sci. Tech., 13,* 255–261 (1981).

INDEX